Subjugate the Earth

SUBJUGATE THE EARTH

The Beginning and End of Human Domination of Nature

Philipp Blom

Translated by Wieland Hoban

polity

Originally published in German as *Die Unterwerfung: Anfang und Ende der menschlichen Herrschaft über die Natur* © Carl Hanser Verlag GmbH & Co. KG, Munich, 2022

This English edition © Polity Press, 2025

The translation of this book was supported by a grant from the Goethe-Institut

Polity Press
65 Bridge Street
Cambridge CB2 1UR, UK

Polity Press
111 River Street
Hoboken, NJ 07030, USA

ISBN-13: 978-1-5095-6132-2 – hardback

A catalogue record for this book is available from the British Library.

Library of Congress Control Number: 2024935053

Typeset in 11.5 on 14 Adobe Garamond
by Fakenham Prepress Solutions, Fakenham, Norfolk NR21 8NL
Printed and bound in Great Britain by CPI Group (UK) Ltd, Croydon

The publisher has used its best endeavours to ensure that the URLs for external websites referred to in this book are correct and active at the time of going to press. However, the publisher has no responsibility for the websites and can make no guarantee that a site will remain live or that the content is or will remain appropriate.

Every effort has been made to trace all copyright holders, but if any have been overlooked the publisher will be pleased to include any necessary credits in any subsequent reprint or edition.

For further information on Polity, visit our website:
politybooks.com

Contents

Acknowledgements

Researching and writing this book was an adventure, an incredible challenge and a fascinating journey. Some of my fellow travellers were especially supportive, and I thank the following people for good conversations, for listening patiently during long excursions, posing astute questions, making suggestions and answering sometimes naive questions: Thomas Angerer, Gertraud Auer-d'Olmo, Tina Breckwoldt, Lothar von Falkenhausen, Michael Ignatieff, Ivan Krastev, Geert Mak, Brian Van Norden, Hannes Benedetto Pircher, Shalini Randeria, Alexa Sekyra, Richard Sennett and Heike Silbermann.

Tobias Heyl accompanied this book from the very first idea, helped shape it and edited it wisely; Sebastian Ritscher moved it along with great enthusiasm and infinite patience. Marie Klinger helped with quotations, sources and important questions.

Amid the silence of the pandemic, more than ever, my wife Veronica was my daily, most important conversational partner.

Thanks to all of you, my fluttering ideas were able to reach the ground and assume a form.

Vienna, June 2022

Illustrations

Up into the Air

A moment of greatest heroism in the face of death, of blind faith, when he runs towards the cliff and leaps into the air, trusting in wings that feel so alien and rigid, until the wind seizes them, bearing them aloft all at once like mere feathers. Then he flies, he rises, racing with flapping wings into the summer air above the island. He sees the houses below, the trees and even the mountains growing ever smaller, the glittering sea extends to the horizon, to the white everywhere. He feels his strength, lifting himself higher with every stroke of his wings and ever farther. He sees his father flying beneath him, who would never have the courage to raise himself so far into the heavens like a god, the ruler of the islands and the sea. But Icarus hears the pulsing blood pounding in his ears and coursing through his bulging veins, he feels every tensing of his muscles, he feels the warm air rushing around him like a flowing embrace, the breath of an unknown goddess. Icarus swings farther aloft, farther than the seagulls and geese, higher than the boldest eagles fly. He has done it. He has escaped from the world down below. Its tyrannical laws. From now on, he will write his own laws. From now on, he will live like a king: sublime, free and ready for any challenge. He will rule over all of this, these little patches of land that look from up here as if a bird had dropped them in flight, as if the gods had rolled dice with the islands to win a prize. In a moment he will reach the clouds, which fly freely and untameably through the sky; in a moment he will be able to touch them and wrest their secret from them.

Only a few more beats of his wings, then he will be there.

Prologue
Buy Me a Cloud

Look at the sky, at its infinity, and before that, at the high-domed tumult of the clouds. Whatever lies on the strip of land below – an alpine panorama, the daily traffic jam on Sunset Boulevard, an industrial ruin, storm-lashed oceans, cornfields or glittering skyscrapers – up there the wind blows freely, and thoughts must be equally free in ever new forms. It must be the last bastion of wildness.

Painters have always been infatuated with clouds, with their stormy metamorphoses, with the sensuality of their forms, the play of light and shade and the dramatic shifts of mood that take over when the sun suddenly disappears, or when it breaks through the looming leaden masses like a revelation.

The greatest cloud virtuosos were the Dutchmen around the middle of the seventeenth century who began to see their own moods echoed in the fragmentation and poetry of celestial landscapes, if only because the terrestrial one had little to offer them: barely a hill, let alone dramatic peaks or canyons, majestic rivers or panoramas. Here everything was damp and small, a brownish hue with some grey, with little that stood out – no ancient ruins or other sources of sublime frisson. The people there were farmers or herring fishers. The land was a line on the horizon, interrupted only by a few trees or a row of windmills. Large parts of this landscape were created by human hands: not only the fields, whose edges were drawn with a ruler, but also the canals and towns, the very land itself, which engineers, dike wardens and the hard work of anonymous arms had wrested from the North Sea. 'God created the Earth', an old saying goes, 'and the Dutch created their own country.' They were not lacking in self-confidence.

But the painters were looking for more than delineated production units for market gardens and dairy pastures. Their patrons, the patricians of Amsterdam and other trade centres, demanded visual representations of their approach to life and their ideas. They were strict Protestants who believed that they were directly accountable to God. Without confession

1 Jacob von Ruisdael, *Wheat Field*, c.1670, oil on canvas, 100 × 130.2 cm.
Metropolitan Museum of Art, New York, estate of Benjamin Altman, 1913.
Accession number: 14.40623

or absolution, they were thrown back entirely on their conscience. The artists of the time projected this drama onto nature. Canvases showing a farmhouse or a little wood provide the stage for psychological dramas in which the cloud masses represent the storm of emotions and the inner struggles.

In the sky, Rembrandt, Ruisdael and their colleagues recognized the last wilderness of a world that had been heaped up, delineated and cut into strips. The sea, that eternal provider and eternal foe of all coastal peoples, represented a nature that could not be subdued and whose force one had to respect if one valued one's life. But the sea was always also a source of fish and merchandise, of work and careers. However much they respected it, people had a pragmatic relationship with the North Sea. The sky was the last space in which the storms of the soul could be depicted.

*

1 July 2021: the hundredth anniversary of the Communist Party of China. A guard of honour marches in front of 70,000 invited guests in uniform and 56 loaded pieces of artillery across Tiananmen Square, and through an enormous gate crowned with the dates 1921 and 2021 as well as a hammer and sickle in gold. The soldiers move with the discipline of a single body; every corner is as straight as a ruler, the metal of their guns sparkles in the sun and their eyes stare rigidly ahead, into a glorious future. While the national flag is hoisted, the cannons fire a 100-gun salute. The Communist Youth and the Young Pioneers enthusiastically pay tribute to the party in front of a gigantic portrait of Mao Zedong. The youngsters have little earphones in their left ears so that they can chant the choruses and party anthems in perfect synchronicity; nothing is left to chance. Helicopters fly above the square in formation, presenting the number 100.

Some distance away from this ceremony and the omnipresent posters, banners and neon signs for the party jubilee, the city's normal, chaotic life continues. The oppressive smog that usually makes it hard to breathe has cleared somewhat. It is a welcome side effect of the festivities for many people in Beijing: the sky is a radiant blue, and although photographs from this day will show a yellowish-grey vapour above the houses, the visibility and air quality are substantially better than on other days, because factories in the vicinity of Beijing with particularly dirty emissions had to halt production a few days before the ceremony.

International scientists found another reason for the fine weather on that festive day, however. The government had employed a technology in which they had invested huge amounts of money in the last few years: *cloud seeding*. Here silver iodide or other chemicals are sprayed on clouds by aeroplanes in order to stimulate the production of droplets, and thus provoke the shedding of moisture by the clouds in a desired location. Thanks to the artificial rain on the previous day, the air was clean and the sky above Tiananmen Square was almost blue. Cloud seeding had also ensured attractive television footage of the 2008 Olympic Games.

According to official Chinese figures, more than 200 billion cubic metres of rainfall had been artificially stimulated between 2012 and 2017 alone, and artillery shells filled with iodine had prevented massive damage from hailstones in 2019. The aim is to expand changes of weather through cloud seeding until one can cover a surface one and a half times the size of India, in order to secure agricultural production quotas and propaganda events.[1]

*

'I, Noa Jansma, sell clouds', a young Dutch artist announces on her website. She explains her project in business language:

1. Prospecting: clouds become my property. Following Jean-Jacques Rousseau's theory of occupation, I take control of it by drawing a boundary around it before someone else. I have trained Artificial Intelligence to do it for me.
2. UR (Unique Registration): according to John Locke's theory of labour, people must interact with the clouds to make them their property. I have built an installation in which people can lie on the grass and gaze at projections of clouds that drift past. The clouds are priced (in €) according to their attributes and are given a QR code. When viewers scan this QR code with their telephones, they enter the world of virtual speculation. As part of their interaction, they share their data (a selfie and their name) with the cloud and receive a certificate.
3. The US (Universal System): after payment, the owners receive a certificate that is also archived in an online land registry. The purchased clouds float in virtual space with the purchase prices. Inspired by capitalist market forces, larger clouds in the registry can eat up smaller ones and grow at their expense.[2]

The pandemic forced Jansma's project to mutate into an online event. Nonetheless, she believes that Buycloud clearly has great potential in the face of disaster: 'New studies predict that with rising emissions, there will soon be no more cumulus clouds. This will cause a rise in temperature by 8° Celsius – catastrophic for the planet, but excellent for the cloud market. The purchase of a cloud becomes a poetic, but stable investment.'

The laughter occasionally sticks in the investors' throats, but the artist plans to take her ideas a step further. Her inspiration came from the history of European conquests of other continents, she explains: 'When in the 15th century the Western "explorers" went to what-we-now-call-America, they told the native people they wanted to buy their land.

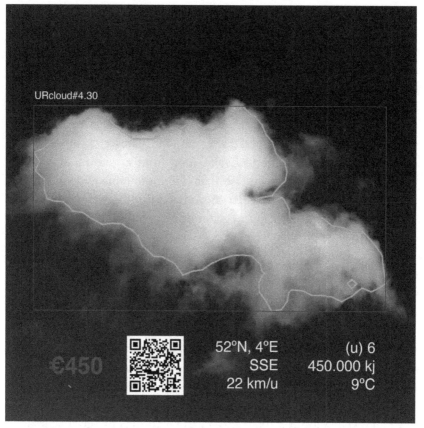

URcloud#4.30

€450

52°N, 4°E   (u) 6
SSE   450.000 kj
22 km/u   9°C

2 Noa Jansma's project Buycloud. Source: https://noajansma.com/buycloud

The natives were confused; "Their land?" "To buy it?" Their vocabulary did not have a word or understanding of ownership over natural phenomena.'[3] As the last remaining not-yet-colonized phenomenon, the clouds are waiting to finally be marketed globally.

*

Clouds – the last untamed part of nature? They are indeed, these ever-changing formations – but only in our imagination. Their growth has long been accelerated by global heating; they are observed, classified, tracked, analysed, chemically manipulated and, in more than one art project, awarded prizes and turned into objects of speculation: future options on the crop yields of individual agricultural commodities, and thus bets on the weather over the harvest period, have long since become normal. One can make a great deal of money from clouds.

Anyone who spends enough time gazing at a landscape – or, rather, a skyscape or cloudscape – of cumulus clouds, a field of finest cirrus in the light of the setting sun or a leaden impending storm front cannot help being hypnotically sucked in by their inexhaustibly inventive variations on a theme. Faces and figures appear, dragons fight with other wonderful creatures, menacing rock faces tower up, sunbeams cut through dark walls or illuminate a scene like something from a Baroque opera. No landscape can be more grandiose than the mountains and canyons of these looming chimeras. As when one looks at flowing water, at breaking waves or a fire, one's consciousness can be completely swept away by the current and even dissolve in it.

The anarchic, intangible, ever-changing nature of clouds has enabled them to elude human control for so long. They have always belonged to the gods, who could crumple them or banish them from the sky at will, or use them to conceal themselves and hurl their thunderbolts.

But now, in a time when smart entrepreneurs and self-appointed visionaries have long been planning simply to abandon our planet, where humanity has behaved in recent times much as a rock band behaves in a hotel suite, and to bring their destructive instincts and ownership claims to other parts of the galaxy with a cosmic Noah's Ark, the space of clouds has long been colonized too. Only in those corners of the imagination not yet usurped or numbed by commercial interests can the clouds still paint their swelling and fading sorceries into the sky – a reminder that everything which is part of nature is in a state of constant flux, impossible to pin down.

The tamed land beneath the clouds and the lust for ever new conquests in the stratosphere are expressions of a collective delusion, the completely unfettered idea that man (the masculine form is intentional here) is outside and above nature and can – indeed must – subjugate it. This conception of humans deems them superior to animals and other living beings, and sees nature as the backdrop to its own ambitions and a warehouse for natural resources. From this privileged position, it sets about subjugating the world entirely to its will.

This ambition is accompanied by the fluttering of a Faustian madness. At the same time, however, this delusion of controlling nature is so ubiquitous and all-pervasive that it is difficult to gain the necessary distance to see it with all its grotesque and fascinating faces, masks and grimaces – which, after all, are also invisible in clouds until one steps out of them and views them from afar.

The subjugation of nature has long since become a global practice. In societies that like to think of themselves as enlightened, and often look back on a Christian tradition, this delusion has especially deep roots in the understanding of nature and conception of humans. It is passed on in families and schools, appears as a pattern in stories, films and video games, as well as laws, remarks and even jokes, whereby the social world presents itself to individuals as holding the same points of reference.

This subjugation shapes the approach to the world and the self-image of many societies that invoke a shared legacy. From their perspective, history appears as the expansion of civilization and the development of progress, which finds its highest expression – whether by coincidence or providence – in their own way of life, or a very similar one. The rise from nomadism to agriculture, urban cultures, text and money, the wheel and railways, human rights, liberal democracies and global markets seems to be advancing with unstoppable momentum.

That, at least, is how observers in the so-called West have described it since the collapse of the Soviet Union, but history has taken several other turns simultaneously. The eschatology of liberal democracies and liberal markets has been replaced on the one hand by the techno-future of Silicon Valley, which clothes the same old longing in new images and enacts it as transhumanism, colonization of distant planets or the reign of Artificial Intelligence.

In other fields, this narrative has been thwarted by reality, from the climate catastrophe to the violent opening of post-imperial wounds and humiliations from the Middle East to Ukraine. Beyond these obvious conflicts, the disregard for natural systems and the resulting collapse of biodiversity is causing a race towards predictable disaster. Instead of a heavenly Jerusalem, what looms in the middle distance is Sodom and Gomorrah.

The tamed and dominated land, the subjugated planet, is proving to be overtaxed by so much wilful and sudden manipulation. Organic connections that developed over millions of years and were stored in the Earth's memory were blown back into the atmosphere within a few decades: their energy fuelled the rapid rise of a species to unimagined power.

From the perspective of ecological systems, however, this rise has a price: finely attuned life cycles are collapsing, the chemical composition and temperatures of oceans and the atmosphere are changing, ocean currents and jet streams are shifting direction, the polar ice is melting, rainforests are disappearing, sea levels are rising and biodiversity is collapsing. The heavenly Jerusalem is still uninhabited, yet already resembles a damp cellar.

These natural processes are taking place as predicted by scientists, except considerably more quickly than in many models. We must therefore be prepared for the next stages of global heating to follow much as calculated, but the potential for repression, denial and political instrumentalization is so enormous that this understandable and observable truth cannot prevail by itself.

And so the disaster is unfolding before everyone's eyes. But *Homo sapiens* is not an especially important organism, and will only have a temporary effect on the fate of its home planet; microbes reigned before humans and will do so after them, and mammals are little more than carrier organisms for them. *Homo sapiens*, of course – and this thought is not without a certain comedy on the evolutionary stage – sees itself as the centre, the measure of all things, the ruler of nature. It actually believes that all living creatures fall into the dust before its incomparable majesty.

In the cold light of day, however, one recognizes *Homo sapiens* as a primate that hopelessly overestimates itself, an inessential part in a system of systems known in Western tradition as 'nature', a biological newcomer that currently seems to be going through the cycle of all innovative species: maximum expansion, degrading of resources and finally collapse. The Roman Empire followed the same path.

The subjugation of nature plays a key part in this unfolding drama, albeit perhaps not the expected one. It has long become part of the fabric in which our societies think and act. It is taken for granted as part of human life, yet its success was never certain; its career has taken a more adventurous course than those of many fictional heroes. The idea of subjugation established itself over centuries in a very restricted geographical and cultural environment before embarking on a new, infinitely more powerful life. It was brought into the world with the ships, the books and the cannons of Europeans; the Enlightenment thinkers declared absolute control over nature to be humanity's noblest task; scientists and engineers seemingly took giant steps towards a glorious future; capitalists and communists alike elevated it to their *raison d'état* and literally declared war on nature.

In this book I will attempt to trace the astonishing history of this delusional idea, from its birth in the dawn of documented civilization to its death in the course of the climate catastrophe.

*

Outside of the 'Western' tradition, the picture is an entirely different one. Scarcely any other society whose myths and stories have been handed down and explored to this day views humans as rulers over nature, superior to the vermin at their feet, destined to subjugate it and complete history.

In Chinese intellectual traditions, for example, the way, Tao, dictates how and where nature flows, and that humans must learn to recognize this way and respect balance (as we will see later, however, this does not occur as idyllically as it initially seems). The Aztecs saw themselves as slaves of tyrannical and incompetent gods that they encountered in all natural phenomena, and which could only be placated with exaggeratedly bloodthirsty human sacrifices.

Aboriginal Australians understand themselves as wanderers along the dream paths of their ancestors, which connect them intimately to their land and form a spiritual geography. The Jívaro people in Ecuador know that they are a people of raiders who live at war with nature and take, either by force or by cunning, whatever they can capture from their omnipresent foe. For the Māori of New Zealand and their Polynesian ancestors, the natural world is full of things and places that are *tăpuu* – taboo – for everyone or only for certain people, meaning that they must not be touched, eaten or entered.

In the Japanese Shinto tradition, the highest aesthetic perfection and the greatest wisdom lie in a meditative identification with natural and transient forms and processes. Members of the San people in Botswana and Namibia know that they are related to animals and trees, and that their ancestors may live in stones or even the wind. It is easy to deride such conceptions as poetic naivety, but cultures like the San have managed over several millennia to live in a comparatively stable relationship with their natural surroundings. The Western model has reached its limits within a few centuries, if not decades.

These conceptions of the world (which are merely a few arbitrarily chosen examples) differ greatly from one another and convey very different ideas of humans and patterns of action. They came about in cultures with widely varying degrees of technological development and social complexity, under very different climatic conditions and in reaction to challenges of different kinds. What they have in common, however, is that they perceive humans as part of a closed system.

Many traditions grant humans a special status of sorts, as described by Dipesh Chakrabarty,[4] but none of these many conceptions of the world involves the insane and breathtakingly narcissistic idea that humans stand above nature and can force not simply other humans and territories, but even nature itself, to submit to them, whether through prayers or through technological arsenals and scientific penetration of the last secrets of the cosmos.

For a long time, this idea was but one of many, and the delusion of subjugating nature was limited to the ambitious fantasies of various monks and scholars in Europe, a part of the world that had descended into anarchy after the collapse of the Roman Empire. Other cultures with other ideas about the world – and no doubt also other collective delusions – developed on other continents. Some societies and their stories existed largely isolated from one another, while others were in constant exchange through migration, trade and war. None of these cultural approaches to the world, however, succeeded in establishing itself on a global scale.

The fifteenth century saw the emergence of a rapid historical dynamic that destroyed this equilibrium. Within a few generations, the narrative of the control and subjugation of nature was globalized, introduced and spread by colonial powers, adapted and often intensified by rebels and freedom fighters, and furthered, praised and executed by churches, communists and capitalists. In the process, other approaches to the world were branded backward and fought, while the gospel of the scientific control of nature in the service of humans, business and progress was hammered into many millions of heads and enforced with armoured brigades if necessary.

Today, this delusion is so endemic and deeply rooted, having reached the furthest corners of our consciousness and our conception of humans, that it is literally impossible for many people to imagine the world from a different perspective. The history of this unique delusion offers one way to gain a critical distance from the idea, which in many ways forms the matrix of the Western approach to nature.

It therefore seems wisest not to treat the idea of subjugating nature as an entomologist would, impaling and classifying it, but rather to describe the entire process of its genesis and observe how it develops, infects new minds and collectives, fights for survival, changes and triumphs – from its beginnings in Mesopotamia to global dominance and gradual death. This collapse gives rise to a philosophical revolution greater than the Copernican one: the radical rediscovery of humans as part of nature. That intellectual adventure will return in part III of this history.

Humans as part of nature come into being when the history of the domination of nature is turned on its head (or, as Marx would say, from its head to its feet). Rather than viewing *Homo sapiens* as the master of creation, it is also possible to understand it as an animal entangled in all sorts of contexts, a nodal point in an infinitely complex tapestry of changing states, a being with less power and freedom of will than it flatteringly ascribes to itself.

From this perspective, then, who is really acting? How important a part of this complex picture are the stories that societies send onto the stage of their collective and individual inner theatres, and which are meant to guide their actions? Can collective ideas and stories play an active part in history, or are they merely passive figments of the imagination? In other words, do humans act more as free individuals or more as part of a collective attunement, a common cultural horizon, driven by the drama of their inner theatre?

Perhaps it is interesting to see the delusion of the domination of nature too, and with it every collective delusion, every story a community tells itself, as an actor – not a biological one, but lifelike nonetheless – that carves out a path for itself with a certain intentionality and creativity, adapting and changing and finding strategies to expand further and infect more minds like a virus, and hence like evolution itself. Subjugation thus emerges as an evolutionary dynamic that uses humans, just as fungi and countless microbes do in the great dance of entanglements and dependencies that we call 'life'. Delusion as an actor: it is this quasi-evolutionary perspective that creates the necessary analytical distance to tell the story of that delusion in the first place.

*

In all these reflections on the relationship between humans and nature, it is the latter that plays the passive part, and I will still refer to it by that name, even though these two concepts will dissolve in the course of my deliberations. The difficulty of reflection already lies in this one word, 'nature', whose meaning one would imagine to be immediately clear – yet doubts arise as soon as one begins to interrogate it, and no one knows what the person they are speaking to actually means by it.

To open up the semantic horizon a little and allow the complexity of this word to appear in a momentary flash, let us remember that the word 'nature' has always transported a difference. Nature is opposed to culture; the one defines the opposite of the other, yet they simultaneously depend on each other. Bruno Latour compares them to 'Siamese twins who continue to hug or hit each other without ceasing to belong to the same body.'[5]

The hierarchy between culture and nature is viewed in different ways depending on the ideological disposition. Nature is untouched and emerges of its own accord (or through divine intervention), while culture is made by humans and constitutes their true destiny. Humans stand between nature and culture. Their historical mission lies in the emancipation from nature and the creation of a higher culture, the foundation of their freedom and their deliverance from earthly shackles.

This somewhat exaggerated narrative is mirrored in an artistic genre that came to life in a period when the relationship between humans and nature changed radically: the still life, which became especially popular in the Netherlands during the seventeenth century. A classical still life – a painting with a bouquet of flowers or a plate of fruit, or cooking ingredients including game and fish – is never the representation of a naturally occurring found scene, but rather a careful arrangement of various elements based on a moral order. And a still life is never alive. The French term for it is *nature morte*, 'dead nature'.

A still life not only ordered nature; it lent it moral content, transforming natural objects such as flowers or fruits into mere ciphers of a divine order. Every bouquet consisted of cut flowers whose death was inscribed in their beauty, of a fruit at the height of its ripeness and about to become rotten, with the first flies already circling it. A candle would soon burn down; a blossom quickly dries; a glass soon empties; a flute is only brought to brief, melodic life by a transient breath – and the frequent skulls, account books (the Dutch were a nation of merchants) or religious treatises require no interpretation. Nature became a moral spectacle, a space in which to stage human mortality and a longing for transcendence.

This thought movement, Latour states, leads to a structural schizophrenia: 'It is nevertheless this unwarranted generalization that gave rise to the strange opinion that has made it possible to deanimate one sector of the world, and to overanimate another sector, deemed to be subjective, conscious and free.'[6] The experiential continuum of nature/culture in which human consciousness exists is split into an individual, subjective, 'overanimated' culture and its shadow, a lifeless, objectified nature. Each conditions the other.

On the one hand, distant 'nature' becomes a mute resource and economic externality; on the other hand, it becomes a still life, a landscape, touristic décor, kitsch. The rest is the historical revenge of Jean-Jacques Rousseau: in a society that has emancipated itself from all natural rhythms, foodstuffs and charms, and increasingly shifts its experiences into a sphere of digital simulacra, authentic and untouched nature once and for all becomes a site of longing, even though such a nature has not existed on this planet at least since the proclamation of the Anthropocene.

At least since Rousseau, the cultural backlash against the artificiality of culture has consisted in a withdrawal to paradise, to childlike innocence and harmony with nature, to an idyll. This is at best dangerous anarchic romanticism, but usually mere conceptual kitsch. There is no return or stasis in nature or in history – no still, neutral place of contemplation from which the world can be described objectively. The mere fact that all thinking occurs in and through ageing, desirous, sick, fearful, ever-changing bodies and experiential horizons renders such a historical abstraction impossible.

Western history offers an entire panoply of tensions and positions between hyperanimated culture and passive nature, between extreme separation from nature and the longing to return to its bosom. At the same time, as the anthropologist Philippe Descola argues, the way in which the modern occident presents nature is 'by no means widely shared'.[7] In many regions on the planet, humans and non-humans are not viewed as fundamentally separate, he explains. They inhabit the same 'ontological niche', have the same needs, are related to one another, are interconnected by the same histories and are fully fledged individuals with their own forms of reason, morality and society.

This separation of ontological niches into Western people and their culture, on the one hand, and what they call 'nature', on the other, is never complete – indeed, one part of this book is devoted to the archaeology of intellectual resistance to it – but it made the culture of subjugation possible, and shaped it, by turning the organisms with which humans share this planet into a *nature morte*.

Wherever the various voices and positions are located in the continuum between ecstatic dissolution and total objectification, they all share the complicated, contradictory history of the concept from which they proceed. In the following, it is important to take the difficult biography of this concept into account while reading and thinking whenever the word 'nature' appears, seemingly innocently, in various contexts and meanings, repeatedly eluding any clear definition.

In one meaning, however, nature has now returned on a massive scale to the lives of many millions of people. The coronavirus pandemic made it drastically clear how arbitrary and costly this separation has become, how vulnerable humans are, how directly they are part of nature, interconnected and entangled in biological, economic, political and social contexts beyond their control, even beyond their knowledge. It is a pandemic that was probably caused by human interventions in nature, and will probably be ended by the inventive spirit of humans.

Yet, even now, the virus has changed our perceptions and instincts, modified how we feel about our bodies and altered our working practices, family dynamics and social rituals. It has increased social differences and exposed governments; it has reinforced people's trust in science in some countries and further eroded it in others; it has divided societies, burdened countless people mentally and financially; it has led to new medical breakthroughs and unprecedented interventions of the state; and it has supplied a new vocabulary for old debates. However long the global emergency caused by it lasts, we will return to a different world.

If a biological pandemic can have such a profound effect on the thinking and behaviour of millions of people within a few months, regardless of whether they were physically infected or not, what about a delusional idea whose infectious power has been afflicting societies time and again for millennia? And what will come after the pandemic? Something always comes afterwards.

I

MYTH

The World on a Vase

3 Stone vase with relief, Mesopotamia, Uruk period, c.3200–2900 BCE.
Iraq Museum, Baghdad

This is the world and its order. A mighty cylinder of pale, yellowish-grey alabaster the size of a ten-year-old child, with horizontal bands of figures winding around the vessel (figure 3).

At the bottom, directly above the foot, one sees rippling waves, the waters of the Tigris and Euphrates, which turn the arid plains into fertile fields; the waters on the coast, where the freshwater and saltwater of a god and a goddess had mingled in an act of cosmic conception to create the known world; the sparkling canals running between the fields and gardens in a dense web. Everything is based on water, which, as in the *Epic of Gilgamesh*, creates, nourishes and surrounds the world.

The next band shows ears of corn and reeds, cultivated and wild plants of this coastal landscape, and directly above them is a band with sheep and proudly horned rams following one another in a seemingly endless procession.

On the next level, a long row of men are carrying jugs of oil or beer and bowls full of fruit to offer up their harvest at the temple. They are all frozen in the same gesture, in profile with almond-shaped eyes, striking noses and bald heads; their angled arms hold their burdens in front of them, their bodies are soft and strongly built, as if to show that their master is rich and they have enough to eat; their legs are opened, the genitals clearly visible.

A wide band separates these inexhaustible carriers from the uppermost depiction, which is almost at eye level for the observer. A naked man, praying, offers the goddess a basket of fruit. The two reed bundles, whose tips are rolled up in the shape of a ring, identify her as a powerful deity.

This field depicts a very special ritual: sacrifice and sacred wedding. The ruler marries Inanna, patron goddess of the city and mistress of heaven, goddess of fleshly love and war, of justice and power. We cannot be certain how this ritual was performed. Perhaps the king consummated marriage with the high priestess as a proxy, but just as the priestess is no longer an earthly woman during this coitus, so too the king is at once the body and proxy of Dumuzi, the divine consort of Inanna. Thus, the rulers performed an annual ritual to ensure the fertility of the land.

Nature – inanimate, vegetal and animal – occupies the lowest levels of the pyramid, whose highest point is the sacrifice at the temple, the mystical wedding, as later Christian authors would call that moment in which worldly life joins with otherworldly life to guarantee the order of the world.

Above the levels of the natural world are the slaves and low-ranking humans (although those depicted here seem to be symbolically undressed priests). Only on the highest level are most figures (except the priest) wearing ritual clothing. The king is taken as the groom to his bride, led by a broad sash (unfortunately, this fragment has been lost). The priest hands over a bridal offering. The bride stands before the entrance to her temple and storehouse, part of the temple complex, just as the priests and administrators belong to the same literate, mathematically educated class. The storage of the grain for meagre years, taxation of fields and fixing of prices formed the basis for the temple elite's power.

In the temple, there are two statues of gods, several offerings – and a pair of vases that look astonishingly similarly to the Uruk vase. Archaeologists assume that this masterpiece from the temple of Inanna had a similar partner piece and both stood in the temple, so the two vessels in this depiction become part of the story they tell: an infinite self-referential mirroring.

This play of references is no coincidence, for the object itself becomes a game. The ritual of marriage between the ruler/Dumuzi and the goddess (represented by a priestess or temple courtesan) was celebrated annually. The sacred wedding also recalled the fate of the god Dumuzi, who had to spend half the year in the underworld and be reborn every year, just like the plants. It guaranteed the continuity of this cycle and simultaneously explains the simple, conical form of the vessel, since the vase can also be thought of as a gigantic cylinder seal which, rolled out into infinity, symbolizes an ever-returning cycle of marriage and harvest, supported by the eternal hierarchy of the divine order.

The Uruk vase is not only for viewing, but also holds its message in its form and the way in which it makes itself part of its own story. It portrays a world in which humans have subjugated the earth and are themselves subjects of the gods – a world in which everything follows a divine order. This order, however, contains an ambivalent element: humans are the only creatures that appear on the vase on two different levels; half animal and half divine, they inhabit an intermediate realm.

Clearly, this dual nature created a tension that was already difficult for the Mesopotamians to bear. People told each other stories about this, as they do about all great tensions, ruptures and fears. One of these stories, about a great king who was two-thirds divine and one-third human, and who set out to subjugate nature and death itself, became one of the central narratives of Uruk. It was recorded on twelve clay tablets at the start of the second millennium BCE by a scribe and priest named Sîn-lēqi-unninni, based on ancient tradition, in a system of signs developed some 1,800 years previously to facilitate stock-keeping: writing.

Gilgamesh the Hero

He who saw the Deep, the country's foundation,
  [who knew the proper ways,] was wise in all matters!
[Gilgamesh, who] saw the Deep, the country's foundation,
  [who] knew the [proper ways, was] wise in all matters!
[…]
He saw what was secret, discovered what was hidden,
  he brought back a tale of before the Deluge.[1]

These are the opening words of the oldest known story in writing, whose basic characteristics, according to archaeological evidence, extend back to the sixth millennium BCE. And this story provides the first document of the idea of subjugating nature. We must therefore recount it here in a certain amount of detail. Gilgamesh, king of Uruk in Mesopotamia, wishes to make a name for himself with heroic deeds and ultimately fails in his attempt to attain eternal life. Despite his wisdom and strength, he finds himself shipwrecked. This hero is wise and yet foolish, an outstanding ruler and yet a tyrant, cruel and yet sometimes gentle: a contradictory and ambivalent protagonist, like all the great figures of world literature.

In the epic's prologue, Gilgamesh appears as a city-builder who has surrounded Uruk with a great wall that rises imposingly from the plain, an edifice such as the world has never seen before – 'Look at it still today: the outer wall where the cornice runs, it shines with the brilliance of copper; and the inner wall, it has no equal.'[2]

But not everything here is as beautiful as the splendid walls. The king oppresses his people. He forces the young men to be ready day and night to amuse him, and claims the right of the first night (the right to sleep with his female subjects on their wedding night) for all virgins in the city. No one can stop him, and so his desperate subjects turn to the gods that they might tame his excessive appetite and free them from the burden of his tyranny.

The gods hear this complaint and decide to distract the overly powerful king. They create Enkidu, a young man who is Gilgamesh's equal in strength and physique, a hairy individual who lives in the wilderness, far from the walled city, grazing with the gazelles. When the citizens hear of this strange creature, they send the harlot Shamhat to trick him. She meets him at a water-hole he shares with other animals and takes action: 'Shamhat unfastened the cloth of her loins [...] For six days and seven nights Enkidu was erect, as he coupled with Shamhat.'[3]

When Enkidu is finally sated, he finds that the animals he lived among now flee from him. His once-innocent body has been 'defiled' by his new knowledge, by his contact with culture, but his nights with Shamhat have increased his understanding, and he stays with her. Through her, he becomes an inhabitant of the city and grows remote from the wilderness. In the town square of Uruk he encounters Gilgamesh, who is just about to deflower another young woman. Enkidu blocks his path and the two fight until the walls shake, yet neither is able to defeat the other. Thus the opponents become friends.

Gilgamesh's longing for affirmation and fame has not remotely been quenched by his new friendship and his newfound popularity among his subjects and the gods. He decides to head to the Forest of Cedar with Enkidu to kill the forest spirit Humbaba, guardian of the cedars. The advisers of the king and Enkidu try to dissuade him, for Humbaba is protected by the god Enlil and is a fearsome monster. But Gilgamesh rejects their attempts, and Enkidu ultimately agrees to accompany him. After heavy fighting, the two heroes succeed in killing the terrible forest spirit. Gilgamesh fells the enormous cedars and fashions gates for the temple in Nippur out of their wood. They also take Humbaba's severed head with them.

The courage and strength of the two friends even impress the gods. The powerful goddess Ishtar has decided to marry the handsome king. Ishtar is none other than Inanna, the 'mistress of heaven' and goddess of love and war. It is said that she deals cruelly with her lovers when she tires of them. Gilgamesh knows this, and snubs his divine admirer. Enraged and deeply humiliated, the goddess arranges for the Bull of Heaven to be unleashed in order to destroy Gilgamesh and Enkidu. But the two heroes can deal with this challenge too, and kill the bull.

7

Gilgamesh holds an extravagant celebration in Uruk, but the gods are outraged by his blasphemy and arrogance. They resolve that one of the two friends must die and send a fever to Enkidu. Gilgamesh is beside himself with pain and sorrow for his close companion, his second self. He refuses to accept Enkidu's death, and remains with him until a maggot falls from the corpse's nose. Only then is he struck by a sudden realization: 'I shall die, and shall I not then be as Enkidu? Sorrow has entered my heart! I am afraid of death, so I wander the wild.'[4]

The loss of his friend makes the great king aware of his own mortality, and he decides to go and find Uta-napishti, the old man whom the gods have granted immortality. Perhaps he can help Gilgamesh to become immortal? After a long and perilous journey, he reaches a seaside inn at the edge of the world: the first Last Chance Saloon in world literature.

He recounts his heroic deeds to the inn-keeper, but she is unimpressed. When he tells of his fear of death and his sorrow, she gives him some advice that remains as wise now as it was millennia ago:

> 'Gilgamesh, where are you hurrying to? You will never find the life for which you are looking. When the Gods created man they allotted to him death, but life they retained in their own keeping. As for you, Gilgamesh, fill your belly with good things; day and night, night and day, dance and be merry, feast and rejoice. Let your clothes be fresh, bathe yourself in water, cherish the little child that holds your hand, and make your wife happy in your embrace; for this too is the lot of man.'[5]

But the wandering hero is determined to cross the water and find Uta-napishti, and he disregards the wise advice. He finds the ferryman, fells seventy trees to use as bargepoles for the crossing and finally finds the old man.

The immortal Uta-napishti likewise tries to dissuade Gilgamesh from pursuing eternal life. Like a mayfly, man is surrounded for a short while by the riches of the world, only to disappear suddenly: 'Man is snapped off like a reed in a canebrake! The comely young man, the pretty young woman, all [too soon in] their [prime] Death abducts them!'[6]

The old man tells him his own story. Long ago, the gods decided to wipe out the town of Shuruppak and all its inhabitants with a flood. Only the god Ea refused to participate. He gave Uta-napishti the task of building a boat and loading all the animals onto it so that they would survive the flood. So the old man built the boat and loaded it up with his family, all the animals he could find, purveyors of all the arts, all his possessions and 'the seed of anything that breathes'. Then the gods began their awesome work of destruction: 'For a day the gale winds flattened the country, quickly they blew, and then came the Deluge. Like a battle the cataclysm passed over the people.'[7] Even the gods were seized with fear when they saw their work. They cried out, lamented loudly and regretted their cruelty, for only now did they comprehend that there would be no one to sacrifice to them any more.

The deluge lasted six days and seven nights, then the water withdrew and the boat came to rest on a mountaintop, from where Uta-napishti sent out a dove, a swallow and a raven to search for land. Finally, he made an offering 'on the peak of the mountain', and the gods 'did smell the savour sweet, the gods gathered like flies' to still their hunger.

After telling this story, Uta-napishti decides with his wife to put their implacable visitor to a test. If Gilgamesh can stay awake for seven days and nights, the old man will convene the council of gods to decide whether to grant him immortality. The hero agrees, but is so tired that he falls asleep immediately. Although he has not passed the test, Uta-napishti's wife takes pity and tells him about a plant that bestows eternal youth. Gilgamesh finds the plant through an adventurous dive to the bottom of the sea.

While the hero bathes in a cool pond on the journey back to civilization, a snake eats the plant and sheds its skin, for it has gained a new life. Gilgamesh risked everything and lost everything, and now he must return to Uruk empty-handed. His last words repeat the opening of the epic: they praise the beauty of the city and its wall, a marvel without parallel.

It is remarkable that the first hero in literary history was already an imperfect man, a seeker who – though two-thirds divine – makes one mistake after another and must suffer for it because he is too arrogant, unheedful of good advice, too proud and too ignorant, because he does not know his place in the world.

Despite being the most ancient hero in all of literature, he does not seem at all alien: he is a person with ambitions and shortcomings that are all too familiar. And this is where we encounter it for the first time: the delusion of subjugation. Gilgamesh, who must defeat everyone in battle, who kills the guardian of the forest and turns the cedars of the gods into timber, the ruler over a city and its gardens who seeks to overcome death itself – this flawed hero is the first bearer of this delusion, of which the myth warns us: you cannot dominate, desecrate, subjugate or disable nature. However far you wander, whatever heroic deeds you perform on your way, it is all in vain against the will of the gods and the laws of fate. In the end, there is no choice but to accept this.

The very first lines mention his greatest achievement, which is connected to his gravest mistake. Gilgamesh 'brought back a tale of before the Deluge' from his encounter with Uta-napishti, knowledge of a harmonic coexistence with the gods given to him by the immortal old man – a knowledge he had not possessed until then.

The epic also explores the ignorance of its hero, for the old knowledge was destroyed through a divine error. When the gods resolved to exterminate humanity with a flood, the god Ea gave old Uta-napishti the assignment to build a boat for himself and his kin, as well as sufficient animals of every species. Uta-napishti was wise; he understood the relationship between humans and gods, but was banished to a place beyond the waters of death, a place inaccessible to humans. It is of this knowledge, retrieved by Gilgamesh, that the first lines of the epic tell.

The *Epic of Gilgamesh* is the story of an ignorant who makes all manner of mistakes because no one remembers how to serve the gods and live in harmony with the Earth they created. Even Enkidu, the child of the forest, is turned by his encounter with culture, with prostitutes and then bread and beer, into a cultural creature from which animals flee.

The View from the Parapet

The city, with its high walls, was Gilgamesh's true legacy, his share of eternity. What did a contemporary of his think when standing on the parapets of the high walls of Uruk and surveying the surroundings all the way to the horizon? What did he think about the rich green of the gardens below or the fields with their sparkling channels in the distance? About the dust-swept landscape beyond civilization, the steppe and the swamp and the mountains? What did he think about the river that nourished them all; what did he think about the steppe grass and clouds, about the flies that tickled his ear and the hot midday sun?

Beyond the parapets, the eye wandered over plantations of date palms followed by fields of barley, flax and sesame, followed by gardens with chickpeas, lentils, beans, onions and trees bearing such fruits as tamarinds and pomegranates: a blossoming landscape. Behind it, as far as the eye could see, pasture in a valley with small farms and villages, a green shimmer over the eternal greyish-brown reaching to the horizon, to the steppe, where Eden once lay.

Here a figure of thought emerged whose effects extend into the present. The steppe as a place of wildness and uncertainty – a hostile place that is waiting to be colonized and civilized, but that can, as wilderness, also oppose culture – reveals itself in the beautiful trope of the biblical Garden of Eden: in the Bible it is the *Gan-ba-Eden*, and in the Avestan language of northern Iran, the linguistic and possibly also cultural origin, it is a *pairi daēza* (paradise), an unfenced garden in the steppe, a protected and shaded orchard in the midst of hostile nature.

The philosophy of gardens fills entire libraries between Japan and England, and asks from the outset whether, instead of subjugation, there could also be a collaborative moulding and evolving of possibilities for shaping nature. The tension between wilderness and taming was always already present in the garden. In medieval Europe, this gave rise to the *hortus conclusus*, the enclosed place in which the virgin and the

11

unicorn live in mystical concord, an organized space that is meant to allegorically represent all orders of creation, and whose plants speak their own symbolic language. The opposition of nature and culture found its expression in this practice, and sometimes in the meditative and vegetative transcendence or negation of these oppositions.

The image of the Garden of Eden does not accompany only Western culture, but it was especially pronounced here and was always characterized by certain motifs that survived over millennia. Uruk, the first culture that could pride itself not quite on subjugating nature, but at least on taming and ordering it through its own diligence and the favour of the gods, was superseded by the Akkadian kings. They continued to use the Sumerian language of Uruk in their rituals, but the introduction of their own tongue was also accompanied by that of a more hierarchical culture.

The social structure of the Sumerian Uruk is difficult to ascertain, since there are temples, but no conclusively identifiable palace districts. This changed under the Akkadians, who set up the first territorial state in southern Mesopotamia around 2300 BCE. It is not only their architecture that displayed a stronger social distinction between rulers and subjects – their thinking was also more vertically structured.

The ceremonial lion hunts were among the high points of royal self-glorification; here, the monarch could style himself as the subjugator of nature and protector of culture. A further innovation of the Akkadian palace city was the royal zoo maintained by several rulers, which presented exotic animals such as elephants, lions and monkeys to an amazed audience: not only domesticated animals were under human rule, but also their cousins from the wilderness.

More time elapsed between the beginnings of Sumerian civilization around 5000 BCE and the Akkadian period around 2300 BCE than separates the present day from ancient Greece, but there were strong cultural continuities that persisted across the various languages and geographical shifts. One of these stable ideas was the view from the parapet over the cultivated landscape amidst the wilderness, the gardens in the desert. Anyone standing there could truly believe, like Gilgamesh, that they were capable of subjugating nature.

Climate and geography favoured this perspective. Mesopotamian urban cultures were the first manifestations of a historical phenomenon that the historian Karl Wittfogel called 'hydraulic societies': communities whose planned and organized crop irrigation allowed them to farm intensively, and which developed together with urban centres, rigid hierarchies and elite military cultures.

Wittfogel's theory of hydraulic societies, which is perhaps overly schematic, has been heavily criticized and partly disproved, but his observation of certain morphological similarities between these urban cultures was valuable nonetheless, especially since such cultures have come about time and again – independently of one another, at different historical moments and on different continents, from China and the Indus Valley to Mesopotamia and Mesoamerica. In Angkor Wat and the Netherlands, canal systems, irrigation and drainage of entire landscapes have proved to be immensely effective instruments in the struggle to dominate a nature experienced as passive or hostile, and for the creation of an intensive and, as Latour would put it, 'hyperanimated' culture.

Field irrigation enabled bigger and more frequent harvests (or, in the case of Mesopotamia, made it possible to have harvests at all). Many people could live in one place, so the production of foodstuffs yielded a surplus and was entrusted to a certain social class. This led to differentiated societies in which merchants, artisans, officials, priests and warriors could all perform their respective tasks.

Taxation of peasants made the rulers of individual cities powerful and allowed them to equip armies – not only to protect their fields, but also to embark on conquering expeditions, for, like Gilgamesh, the Akkadian rulers could only gain glory if they brought riches and loot to the city. At the same time, efficient taxation demanded a functioning system of bookkeeping and administration.

These urban cultures produced an entirely new model of power and social cohesion. The big step towards agriculture had already been taken around 12000 BCE, albeit often in conjunction with a nomadic or semi-nomadic way of life. For a long time, prehistorians debated the social consequences of the agricultural revolution and claimed that it had brought about an abrupt shift from living in small groups of hunter-gatherers to existing as obedient peasant labourers. Recent excavations, however, show that fields and small, flexible settlements often coexisted, even over centuries (just as the Akkadians and Sumerians appear to have lived with and alongside one another for a long time), without fortifications or obvious social hierarchies, without a strong distinction between inside and outside, and hence without a pointed opposition between nature and culture. This was because communities lived, at least for part of the year, in a way that had changed little since the Palaeolithic Age and could be described with the same stories and legends. Walled structures such as Çatalhöyük in present-day Turkey and Knossos in Crete seem to have acted as gathering places for the different communities and their rituals, although recent research suggests that Çatalhöyük was intensively inhabited in certain phases.

In Mesopotamia during the fourth millennium BCE, the picture of a densely populated city had radically changed. Anyone visiting Uruk for the first time who witnessed the bustling streets, the great rituals in the temple, the wealth, the smells and sounds of a city in a world without cities, and around it the gardens, fields and villages, could easily conclude that the Sumerians were capable of subjugating the natural world itself.

The irrigated Mesopotamian fields bear eloquent witness to an administration and social division of labour in which there were master builders, bureaucrats and workers – a system that grew from humble beginnings to incredible complexity and changed the landscape to the same degree as society could reinvent itself, for it led to an even greater concentration of the population, more trade, more wars, more slaves and a society with an even stronger division of labour ruled by an elite of priests and aristocrats.

This produced islands of urban culture whose power was based on water management and which, despite cultural differences, were remarkably similar in some respects. A landscape of irrigated fields and gardens with small farms and villages, transport canals and relatively good roads surrounded a city, which was usually structured concentrically and displayed its social geography in the topography of the dwellings and public spaces: aristocrats' palaces and temple districts; squares for markets and public rituals, but also for executions; ample lodgings for foreigners, but also street shops, brothels, barracks and a city wall.

This organization describes fifteenth-century Tenochtitlan or, three centuries earlier, Angkor Wat in Cambodia, the Mesoamerican city of Tikal or the urban cultures of Mesopotamia.

Landscape and Memory

While exercising due caution about tempting speculations, one can scarcely deny the climatic and geographical influence on agricultural practices and products, livestock and raw materials, and hence also on social structures and their histories. For example, Mesoamerican cultures went through a comparatively weak economic and power-political development, partly because they lacked any pack or riding animals before the Europeans arrived: llamas and alpacas are not suitable for such tasks. Long-distance transport had to be carried out by humans, with only a few routes where canals could be used. A carrier could transport food for 30 kilometres at most before needing more to eat than he could carry.

How much a society influenced its natural surroundings also depended on climatic conditions, however. The Khmer in twelfth-century Angkor Wat or the Aztecs in fifteenth-century Tenochtitlan lived in tropical regions where the organic world is constantly penetrating everything man-made and resolutely trying to take it back. From mould to ants and infinitely inventive plant shoots, nature is incessantly active; in the rainy season, streets become impassable and life retreats indoors as far as possible. In such a climate, the notion that one could force nature to its knees is rather less likely to arise.

The civilizations in Egypt and the Indus Valley lived under different climatic conditions, but both depended on the river, their lifeline, flooding its banks once every year and not only watering their fields, but also providing them with nutrients. As in Cambodia and subtropical Mesoamerica, nature functioned here as a dominant actor and timekeeper, albeit in a nourishing role. An Egyptian would scarcely have imagined dominating nature, since their existence depended on the rhythm of the river, whose deities therefore received offerings. One could open a reservoir if the rains failed to come one year, but no human could control the annual flooding of the Nile or the Indus.

We do not know how equitably and communally organized Harappa in the Indus Valley really was, but Egyptian culture has been sufficiently well explored for us to venture a comparison with Mesopotamia. In their aggressiveness, their systematic slavery and subjugation of other peoples, their taste for monumental architecture, their hierarchical organization and their hunger for power and glory, the pharaohs were every bit the equals of the Mesopotamian rulers, and this indeed made them geopolitical rivals in the second millennium BCE.

And yet there is no Egyptian Gilgamesh, no Egyptian god-man who would subjugate everything. A pharaoh could claim to be descended from the gods and demand to be worshipped like a god; he could build temples and wage glorious wars; but the myths belonged to the gods alone and described the cycle of life and death, the rulers of the different realms. Osiris, the suffering god who had to die and rise from the dead again every year, thus set the rhythm for the year's agricultural cycle of conception, birth, maturation and death, the seemingly eternal breath of annual floods and the fertility they brought.

Uruk lay in a river plain marked by marshland and steppe. There was little rain and the summers were hot; water in jugs could only freeze during the winter. This landscape had genuinely been transformed: date plantations spread out before the city walls along with orderly green fields; nature was tamed. It appeared to have been subjugated.

None of this would have been possible without force, without the exertion of tyrannical power and, even more concretely, without grain, writing and effective taxation. The agricultural historian James C. Scott stresses that this development was only possible in societies that cultivated grain. The conventional narrative describes the development of agriculture as a form of progress that allowed humans to leave behind the insecure, impoverished existence of nomads.

However, analyses of the skeletons of settled and nomadic prehistoric communities in the Fertile Crescent paint a different picture. The bones and teeth of the farmers show that their diet was less rich and less balanced than that of their nomadic relatives. But their bones also displayed more stress fractures as a result of hard labour; the domestication of poultry, cattle and swine exposed them to more pathogens; and they died younger. For these people, then, the change of lifestyle was far from an improvement, but could only have come about under

16

duress, for example through an outside threat such as attacks by other clans or armed gangs. Small villages could not defend themselves, and would therefore have to find protectors, who would demand tribute and service in return.

These results undermine the historical narrative of progress through agriculture and urban cultures. The anthropologist Guillermo Algaze summarizes the new state of research thus: 'Early Near Eastern villages domesticated plants and animals. Uruk's urban institutions, in turn, domesticated humans.'[8]

The most important instrument for this taming was not the sword but taxation, and for this it was essential that the settled farmers, who were subjects from now on, cultivate grain in open fields. The enormous advantage of grain crops from the perspective of the tax collector is easily understood: an animal caught on the hunt, fruits gathered in the forest, wild grains or even vegetables growing in the earth can hardly be monitored. A farmer will rarely say how much he has really harvested, gathered or caught, but a cornfield is in plain view. It has a fixed surface area, a fixed harvest time and a fixed, weighable and countable amount of grain that can be calculated in advance. It is ideally suited to oversight and taxation, as well as transportation and storage. Whoever controls the grain controls power.

Entire states could be built on grain. Voluntarily or forcibly settled farmers could cultivate the fields and be enlisted at other times for work such as building canals, temples or fortifications. The taxes imposed on them in the form of grain and other products required an efficient bureaucracy capable of complex accounting. Cuneiform script was developed in this administrative context as an aid to memory and a tool for documenting stores, debts and credit. This evidently does not apply to all early writing systems – the earliest uses of written characters in China and Mesoamerica were exclusively in ritual contexts.

The society of Uruk still felt the memory of its subjugation in its bones. Its stories and myths revolved around power and tyranny. In *Enuma Eliš*, the creation myth, the father god Apsu decides to murder his entire offspring because the young gods are too noisy and are disturbing his peace. This somewhat frivolous motive leads to a disastrous war between the gods in which, as if by chance, the Earth is created from the corpse of the rebellious water goddess Tiamat, and humanity then from the blood of another god.

According to the myth, the only reason for the gods to create humanity was that the work of cultivating fields and making a living became too much for them, and they wanted to keep humans as slaves who would toil away in the fields in their place and sustain their divine lords with their offerings. *Gilgamesh* likewise deals with power and the abuse thereof, the subjugation of nature and its very laws through the superhuman strength of a hero who, in the end, hopelessly overestimates himself.

Keeping this in mind, it is safe to assume that the Uruk vase would not have posed any problems for that contemporary of Gilgamesh who went from the parapets of the city to the temple to admire the pair of ritual alabaster vessels. The bands on the vase and the figures in them were clearly identifiable: water as the foundation of all existence, the cultivated plants, the neatly marching animals and naked servant figures, then finally the offering in the temple, where humans gave the gods their due and, through this ritual, legitimized their own social order at the same time.

The trained eye, moving upwards from the bottom, reads in this picture that society is a pyramid extending from the lowliest to the divine. Where in this pyramid an individual is born is decided by fate, the will of the gods. One will probably be a farmer, or a slave woman, or a servant or a poor wretch or a prostitute – but that too is the will of the gods, and the will of the gods must never be disobeyed, for they strike back mercilessly with disease and misfortune, defeat and shame. Even the king must sacrifice to the gods above him and defend them against the other gods, their favourites and their mercenaries. The world consists of the rulers and the ruled, and even the rulers are ruled by the imperatives of their status and the implacable dynamics of the circumstances in which they play a part they never wanted. The gods of the Sumerians are the first slaveholders in history. The city's rulers merely conform to their capricious, unyielding will.

'Nature' plays no part in this conception of the world, insofar as neither the Sumerian nor the later Akkadian language spoken and written in Uruk, as well as other urban centres in Mesopotamia, have words for 'nature' or 'culture'. The surviving myths and everyday documents frequently mention sky and earth, land and sea, sun and stars, fields and gardens, plants and animals, yet no great something that grants humans experiences that do not come from other humans, but rather from an original reality – except Eden, that expansive wilderness.

The countless Sumerian and Akkadian clay tablets that have come down to us contain not only epics and literary works, but also a wide range of administrative documents, merchants' letters, handbooks and inventories. The different elements of nature were thus perceived very specifically as long as they were connected to people's interests. Date palms were bred on plantations, and gardening books knew the best times for harvesting along with all other aspects of cultivation; astronomy had advanced far enough to predict solar eclipses and calculate planetary orbits; trade routes through diverse and dangerous landscapes were exploited to introduce such commodities as copper, tin, Chinese ceramics and lapis lazuli; yet, despite this detailed knowledge, nature only ever manifests itself as a threat or a resource. Outside the civilized domain lay the wilderness, the realm of barbarians and evil spirits.

The Free Market of Offerings

We can only know what our Mesopotamian saw on the wall if he speaks about it, and if he were to speak about it in the manner of the many surviving hymns, magic spells and epistolary addresses, he would first of all thank the gods who made such beauty and fertility possible; for the gods, according to his understanding, were genuinely physically present in the world around him, and an exchange took place between them and humans that was akin to exchange in a market, albeit in a forced metaphysical community.

After the embarrassing failure of the flood, which not only drowned most humans and animals, but also brought the gods to the edge of starvation for lack of offerings, humans were in a strong negotiating position in relation to the gods. Each side needed the other, for the gods would go hungry without offerings, and no human could get far without divine protection. The logic worked in a remarkably similar way to the loyalty of today's football fans, who suffer with their team, make great sacrifices and are disappointed time after time, yet can sometimes celebrate unforgettable triumphs. But clay tablets reveal that a Mesopotamian who felt ignored or betrayed by their local god could certainly look for support elsewhere. People wrote requests to their god and, on some occasions, simultaneously threatened him, more or less subtly, with the possibility that they would turn to another deity if he failed to deliver any results. As in the political world, in the battles between empires, kingdoms, city states, armies and families, a Mesopotamian could enter different alliances of convenience with local or general gods and goddesses, be they long-standing or newly adopted from other peoples: a realpolitik of worship.

The gods demanded tribute in the form of offerings, rituals and effusive hymns of praise. They treated individual humans as the king treats a provincial governor, who is left in peace to do as he likes within his jurisdiction as long as he sends enough taxes and war booty to the capital. As long as the sweet smell of burnt offerings rose to the sky, the gods and creators turned their Earth over to the humans. So people can do what they like with the world around them as long as they loyally pay tribute and follow the relevant commandments – or at least maintain the pretence of doing so. Referring to sacred texts, Jean Bottéro writes:

> There was absolutely nothing 'mystical' about Mesopotamian religion. Its gods were considered to be very high 'authorities' [...] upon whom one depended in complete humility, obligated to serve them: they were distant and haughty 'bosses', masters and rulers, and above all not friends! One submitted to them, one feared them, one bowed down and trembled before them: one did not 'love' or 'like' them.[9]

As long as the gods were satisfied, the kings of the city states could focus on their real assignments: expanding the kingdom, keeping rivals at bay or defeating them in a glorious battle, sending out lucrative looting expeditions (a typical political practice in the Bronze Age), erecting monuments to themselves in the form of large-scale construction projects, gathering together the most brilliant craftsmen and artisans and the best builders and engineers in order to immortalize themselves – and then, after working through this long but nonetheless incomplete list, they could take the advice of Gilgamesh's inn-keeper: celebrate every day, honour beauty and go to sleep in the arms of a beloved.

However, the minor provincial ruler, who had to come to terms with the distant central government but, at the local level, could extort taxes, divert treasures into his own pocket and rule with despotic brutality (governors rarely lasted long), is only one side of the relationship between the Mesopotamians and their gods. Although, on the one hand, they had to make the right sacrifices to the gods in the temple, the deities, demons and spirits were constantly around them, inhabiting mountains and rivers and fields, houses and heads, magical amulets and herbs growing at the wayside. The divine was everywhere.

The rituals – whether officially at the temple or in private spaces – were merely the practical expression of the relationship that connected humans to other creations of the gods. That did not prevent either the powerful of the time or the ordinary people (to the extent that one learns about their lives from the documents) from pursuing their own interests robustly. But, like their hero Gilgamesh, they knew that their plans might interfere with those of the gods, and that they would pay a price if they failed to bring the offended deity on side through offerings and other more or less subtle acts of bribery. It was a matter of action and reaction.

Gilgamesh incurred the wrath of the gods when he killed Humbaba and contemptuously spurned Ishtar. It was clear that the gods would exact bitter revenge on him, that they would claim a life – even if it was not his, but that of his best friend Enkidu. That was simply the way of things. The Akkadian word for illness translates as 'hand of god'.

Humans lived through the will and the power of the never-ending conflicts and whims of the gods. These were neither universal (every city had its own protective deity) nor just, all-knowing or necessarily good, but one had to take their interests and influence into account.

It is clear that the myth of Gilgamesh runs through five millennia of Mesopotamian history. Its conflicts, images and figures, however, have asserted themselves in various guises to the present day. Gilgamesh resonates through the Bible, he returns as Odysseus and Parsifal, his egomaniacal rampage anticipates Faust, Prometheus and Orpheus – there is scarcely a story of our time – from Hollywood to Netflix, *Fortnite* and the narrative patterns in the media – that does not still draw on the archetypes of the *Epic of Gilgamesh*.

Gilgamesh was the only one to bring knowledge from before the flood. It seems that historical knowledge is hardly possible before this flood, and whatever survived consists of hard stone or ivory, not the gossamer weaves of age-old stories – even if their threads are possibly still being spun to this day.

Before the Flood

4 Venus of Willendorf (Wachau), *c.*20,000 BCE, oolitic limestone tinted with red
ochre; height: 10.5 cm. Vienna Natural History Museum

She is one of the most famous women in art history, yet we know practically nothing about her. The so-called Venus of Willendorf was found near a village in the Austrian Wachau Valley region. Her age is estimated at over 20,000 years. She is just under 11 cm tall and made from a stone that was brought to the Danube from over 100 kilometres away. Nothing else is known about her.

Who is this woman with the large hips and breasts and the clearly visible vulva? Is she wearing a cap made of little conch shells, as found in other Stone Age graves, or does she have short, curly hair like people from Africa, from where her ancestors may have come not so many generations earlier? The first people to settle in Europe had dark skin. This tiny beauty remains silent, and that is precisely what turns her into such an object of projection.

Generations of researchers have attempted to unlock the meaning of these Stone Age figures, of which some 200 have meanwhile been found – from Siberia to Spain, Romania and as far down as Egypt. What did they mean for humans, and what story do they tell? Can they reveal something about the way people thought at the time? Are they fertility symbols testifying to a more intimate, symbiotic relationship with nature? Are they different goddesses or all representations of the same fertile mother goddess?

These enigmatic idols are among the few material relics from a distant past that might allow us to draw conclusions about the intellectual world and narrated knowledge of people who lived a thousand generations ago. The famous cave paintings seem to come from a world of shamans and animated nature in which animals are ritually invoked or trapped, and in some cases there is a shamanic figure with horns on its head that evidently carries out rituals. However, much of this depends on the interpretation, the condition of the artefacts, the degree of decomposition and often also what it is that researchers hope to discover with these finds, what they read into them and then triumphantly claim to have revealed. The fewer finds there are, the less context there is and the greater the distance between provable facts and plausible-sounding hypotheses.

The existence of female figures, some of them fashioned with great care, gives an indication of the status of women in society, and since the subjugation of nature is a traditionally male-dominated business, these depictions at least open a small window on the mentality of people who lived up to 30,000 years ago.

Women probably enjoyed more respect in Palaeolithic groups of hunter-gatherers than in the villages and cities of later times. In a small group – here comparisons with present-day groups of this kind are actually useful – every hand is important, every skill is precious, and every pair of eyes can save a life. No man has the surplus means to tie several women to himself, and no woman can be kept away from productive, essential work by an excess of infants.

The bead-like decoration on the head of the Venus of Willendorf recalls a finely woven cap made of conch shells found on the head of a female skeleton in a cave in Liguria, separated from the Austrian find by several millennia. This woman had been covered in ochre and buried very carefully, and the same cave complex housed thirteen small Venus statuettes that were fashioned from a harder rock and in a different style, but with the same curves, always referred to by polite archaeologists as 'emphasizing fertility'.

All this points to cultures – not covering large areas, but constituting groups nonetheless – that repeatedly communicated and traded with one another across great distances. How else could the Venus of Willendorf have acquired her shell cap, which was probably a status symbol among the women of the time? And before one knows it, this leads to effusive speculations and flourishing panoramas of happy, equitable, peacefully trading societies in which no evil could be done.

As in other cases, however, not everything about the Venus figurines is as simple as it seems at first glance. Are they really statuettes of the mother goddess, whose cult was gradually replaced by more patriarchal, aggressive conquering societies?

The idea of a Stone Age matriarchy before the spread of agriculture has a long and fascinating tradition. The Swiss scholar Johann Jakob Bachofen already advocated this theory in 1861, in a work with the then-provocative title *Mutterrecht* [Mother Right], which caused a scandal at the time because it proposed that a peaceful, matrilineal advanced civilization was usurped by an aggressive patriarchy. The consequence was that the world not only was crueller than before, but had also eliminated the traces of the murdered civilization, as a killer disposes of the corpse.

Bachofen was a jurist and classicist; his enormous collection of material on early matriarchal societies was based on text sources, which excluded any research on societies existing before the discovery of writing. Nonetheless, the idea resonated after his death, inspiring such disparate personalities as the poet and historian Robert Graves, the psychiatrist and analyst C. G. Jung, the mythographer and archaeologist

Marija Gimbutas, and Joseph Campbell, author of the monumental *The Masks of God*. From nationalist mysticists around 1900 to second-wave feminists, a broad range of thinkers referred to the Swiss professor, whose work had been almost entirely forgotten in his lifetime and was only later rediscovered.

In Search of Lost Matriarchy

The thesis that human societies used to be gentler and that not only were women in charge, but property, succession and offspring were also defined by the female line, does not merely have a certain heretical charm, but actually explains certain peculiarities of ancient history. Great epics such as *Gilgamesh* and later the *Odyssey* and *Iliad*, the Bible and the Indian *Mahābhārata*, suggest that older myths and rituals were pushed out and reinterpreted. These earlier elements dealt with the worship of female deities and came from societies that were more peaceful, less brutal and also matriarchal. According to the theory, the time before the flood belonged to women and was first undermined and then destroyed by patriarchal farmers, after which it was also pushed out of the collective memory and rendered invisible.

Around 1900, publications about Arthur Evans's excavations of the palace of King Minos of Crete gave an enormous boost to the idea of a historical matriarchy. The theses of Bachofen and his students were finally supported by archaeological findings and proved to be correct. In the ruins of Knossos, Evans discovered not only an archaic kingdom that had flourished long before the Greeks, but actually a completely different society. The frescoes and artworks in the palace depict athletic young women leaping with great daring over the backs and lance-like horns of a bull, a powerful high priestess holding a bundle of snakes, like Zeus with his thunderbolts – all the artefacts expressed a culture in which women exercised previously unknown power and had a fearless, sensual presence.

The palace of Knossos was a central place in the European imagination – not only in the Bronze Age, but also in the early twentieth century. During the rise of new technologies, the story of Daedalus and Icarus reminded people of the dangers of human hubris, just as the Minotaur, half-bull and half-man, stood as a warning that unnatural human lust created dangerous monsters. These legends could be interpreted in different ways, but they spoke directly to a society that, between Mary

Shelley's *Frankenstein* and Freud's dangerous unconscious, was fascinated by the dark and destructive side of the bright new world, an original sin denied by civilization that, like the Minotaur, needed to be hidden and rendered harmless, but continued to devour innocent life.

There were even deeper resonances that made Knossos, its rituals and its social world seem meaningful. After the brutal disillusionment of the First World War, the time was ripe for the rediscovery of other societies, other forms of domination and authority than the perversion of patriarchal magnificence that had led to catastrophe. The self-evidently natural sensuality of this Mediterranean spoke to a generation whose own gender roles were becoming shaky. Women's rights activists were demanding new rules for coexistence, neurasthenia was the psychological epidemic of the time, men felt questioned in their sexual identity and sought refuge in masculine rituals, the military and obsessive masculinity. In this context, the vision of an advanced civilization of existential lightness led by women created a powerful antithesis to the rigid male-dominated present with its moustaches and uniforms – to say nothing of the socially enforced hypocrisy in sexual matters.

Another world emerged from beneath the soil of a flower-decked hill in Crete, and with it the history of its annihilation by the ancient Greeks, who had conquered the land and destroyed not only its palaces, but also the stories and memory of the island's culture, which had originally revered the great mother goddess and her lover and companion, the holy bull.

In fact, Evans and other researchers were able to provide evidence through many finds, in an area extending from Europe eastwards to India and Mesopotamia and southwards to Egypt: a horned god (usually a bull aurochs, in the north also a stag or reindeer, or more rarely a ram) appeared in various cultic contexts and depictions whose historical beginnings extended into the deepest prehistorical night. From 15,000-year-old cave paintings in southwestern France to Gilgamesh's heavenly beast, from the antler headdresses of northern England and cylinder seals from the Indus Valley and bull reliefs from Çatalhöyük from the ninth millennium BCE to the horned god of Enkomi in Cyprus and the Apis bull in the Egypt of the first dynasty in the third millennium BCE – there is ample evidence of horned deities and shamans who wore horns for rituals, as well as Mesopotamian rulers who appeared in horned helmets.

The horned god definitely existed. He was powerful: a connection between humans, animals and their spirits, between hunters and their prey. His potency ensured rich harvests.

But then, especially in the Mediterranean region around the eighth century BCE, a new type of society established itself. Agriculture, fixed settlements and equally fixed social hierarchies finally replaced the life of small, more or less mobile communities. They had the more efficient technologies, agriculture fed a larger number of people, and the taxation of harvests by the ruling families was the beginning of states, administrations, armies and property, but also division of labour, palaces, markets and libraries.

These new societies came with their own religious traditions, which favoured hierarchies, and whose myths slowly but surely replaced the memories of the gentler, matriarchal old culture. This process of replacement was all the more effective because the growth of agriculture and the taxation thereof were accompanied by writing; the myths of the conquerors were therefore the first with the capacity to record and spread their version of history.

One can find clear signs in the Bible of this struggle of a part of the priesthood against the horned god and the mother goddess. It is no coincidence that the apostate Israelites dance around a golden calf and are punished for it. In the Levantine religion, the Mesopotamian goddess Ishtar/Inanna was worshipped under the name Asherah, either alone or as the consort of the god Yahweh: a classical pairing for the mythical neighbourhood of the time, like Ishtar and Tammuz, or Isis and Osiris. Hebrew and Aramaic inscriptions from the eighth and seventh centuries BCE mention 'Yahweh and his Asherah' as protective deities that are called upon together.

Evidently this divine couple was also worshipped by the Judeans for a long time. Even King Solomon quite naturally had a shrine to Asherah erected in the temple he built to Yahweh, as reported in the Book of Kings after its destruction by King Hezekiah, then again by King Josiah, for the cult was clearly very persistent. And yet – times had changed. The priesthood in Jerusalem had put their fate in the hands of a single god – one of the countless local deities that also existed in Mesopotamia – who had promised them land and power as long as they followed his laws and respected their holy covenant with him.

If the god of the Judeans was once the consort of the fertile Asherah, their relationship ended in a terribly acrimonious divorce. Yahweh would not tolerate anything within his domain that recalled his former partner, as stated in Deuteronomy 12:2–3: 'Destroy completely all the places on the high mountains and on the hills and under every spreading tree where the nations you are dispossessing worship their gods. Break down their altars, smash their sacred stones and burn their Asherah poles in the fire; cut down the idols of their gods and wipe out their names from those places.' The 'Asherah poles' referred to in Hebrew as *asherim* may have been wooden objects or trees that symbolized Asherah in her shrines, just as they represent the shrine to Inanna on the Uruk vase – perhaps the *axis mundi*, the tree of life, an indispensable element of shamanic rites.

The whole passage in Deuteronomy (known in the Hebrew Bible as Devarim) shows the lord from his most unattractive and violent side. He speaks of his campaign of conquest with his people just as Ashurbanipal would have done, as a commander proud of his greatness; 'his mighty hand, his outstretched arm', with which he destroyed Pharaoh's chariots; and with rival tribes, 'when the earth opened its mouth [...] and swallowed them up with their households, their tents and every living thing that belonged to them'.[10] So he was not the sort of god one would want to pick a fight with.

God promises his people that they will take possession of the land into which he was leading them, ironically adding that it is a land 'in which milk and honey flow'. This god proposes a very simple trade: his people must love him, 'then I will send rain on your land in its season, both autumn and spring rains, so that you may gather in your grain, new wine and oil'.[11] He warns of any breach of contract, any temptation that might cause the Jews to stray from the right path: 'Then the Lord's anger will burn against you, and he will shut the heavens so that it will not rain and the ground will yield no produce, and you will soon perish from the good land the Lord is giving you.'[12]

The Lord promises his own a great kingdom if they refrain from turning towards other gods – the conquest of the entire known world at the time: 'Every place where you set your foot will be yours: your territory will extend from the desert to Lebanon, and from the Euphrates River to the western sea.'[13]

Thus the erasure of the name of Asherah in the conquered territories turns into ethnic cleansing through the destruction of an entire culture.

The astounding results of Arthur Evans's excavations suggested that there must have been similar processes of erasure and forgetting in Crete, and hence the Mediterranean region. According to myth (passed on mainly via the Roman poet Ovid), the sea god Poseidon sent his protégé, King Minos, an especially perfect bull from the sea for him to sacrifice. But Minos kept the bull for himself and sacrificed a different one. Poseidon took revenge by filling Minos' wife, Queen Pasiphaë, with such insatiable lust for the bull that she ordered Daedalus to build a cow-shaped construction that would enable her to mate with the magnificent animal. The result of this mythological mésalliance was the Minotaur, whom Daedalus had to imprison in a labyrinth from which he would never escape. But the Athenians had to send the tyrannical Minoans a tribute in the form of youths and virgins, until Theseus finally defeated the Minotaur with the aid of Ariadne's golden thread.

But what if this story is propaganda spread by the Greek conquerors? It is not difficult to see something else behind the moral of the story of the noble Greek hero and monster-slayer, namely the discredited bull cult and the strong woman at its centre, whose devotion to the sacred and powerful lover of the goddess is presented in the Hellenic retelling as a perversion. The result, a creature half-animal and half-man, is too monstrous to be allowed to live in the sun's light. He devours the young bodies that, in the art of Knossos, took daring leaps over his massive body completely unharmed. Minos himself is the son of Zeus and Europa: the one case in Greek mythology when the father of the gods himself assumed the form of a bull to satisfy his lust.

Setting down old myths in writing was a factor in deciding who had the power to define the narrative in the ancient world, and there can surely be no doubt that after the so-called collapse of the Bronze Age around 1200 BCE, the new cultures that became dominant always sought to control the legacy of their predecessors through their foundation stories and laws, which were now immortalized on stone tablets and stelae.

From here on, however, everything becomes complicated. The written sources reveal that older cultures were suppressed, their rites condemned, their shrines desecrated and their stories retold. Before textualization, however (which took place at very different times in different cultures), there were only isolated artefacts that happened to have been fashioned out of durable materials – individual mosaic pieces, which offer infinite room for speculation and projection in the spaces between them.

The palace of Knossos presented to an amazed world by Arthur Evans probably never served as a palace, but rather a ritual and temporary gathering place for the clans of a particular area. Its architecture, with a labyrinth of small rooms in the interior, would have made it completely unsuitable as a residence, but ideal to provide storerooms for annual celebrations. And the famous frescoes and reliefs on which Evans based his reconstruction of the Minoan religion proved, at least in part, to be a patchwork of completely different depictions, while the architectural reconstructions on site were made of concrete and had more than a touch of Art Deco about them.

The royal palace of Knossos is a well-intentioned historical falsification, and the historical society of prehistoric Crete is still something of a mystery. But one thing is certain: even if Crete was not quite a matriarchal paradise, some of the most impressive graves, which indicate high social status, did belong to priestesses – and the bare-breasted maidens who leapt over bulls were not figments of a Victorian scholar's overheated imagination. So women, or at least some women, genuinely enjoyed a substantially higher status than in later centuries. This is also demonstrated by the burial practices and objects, as well as an analysis of the skeletons.

To reconstruct the shift from an at least partly nomadic way of life to a settled one, prehistoric societies are often compared to surviving communities that seem sufficiently similar in their size and degree of technological development, and whose social structures and climatic challenges have led to particular ways of life, rituals, economic practices, attitudes and narratives.

Such parallels should be treated with the greatest care, however. First of all, it has been thoroughly documented by ethnologists that different ethnic groups living in the same area under very similar conditions, such as the indigenous Amazon peoples, have constructed completely different realities and narratives for themselves to connect their experience with their surroundings and gain meaning from it. In addition, the idea of an essentially ahistorical society that has survived practically unchanged for millennia and preserves authentic myths and traditions is strangely paternalistic, and ignores how indigenous societies confronted with new challenges or traumas can overcome them in resourceful ways and insert them into their 'timeless' narratives as mythical guarantors of an evolving identity.

The Venus figurines, too, ultimately remain unexplained. The French palaeoanthropologist Alain Testart even doubts that they are goddesses. If one wishes to draw parallels to societies of today with similar techno-logical and social structures, he remarks, one should note that such images of women were produced exclusively by men. 'What makes us think that the Neolithic religions were devoted to the cult of the mother goddess? Only these statuettes of naked women, nothing else.'[14]

Only relatively few of the many Neolithic graves belong to women, Testart argues; cult practices are unknown, temples scarcely identifiable as such, and plans of villages tell us nothing about the lives of women in these societies. The figures are certainly not found at excavation sites known as cult locations, nor are they in any way monumental, made from especially precious materials or created in order to impress. Most of them are barely the size of a hand.

So what do the figurines tell us about the status of women, or indeed goddesses? Testart ventures a comparison across millennia. Cultures that remained oral until their colonization and came into little contact with other civilizations offer a way of understanding: 'Nothing is more common, nothing more banal, than these statuettes, which are mostly made of wood and depict women with accentuated breasts, heavily sexualized [...] but no ethnologist or art historian would ever have concluded from this [...] that women were dominant in the respective societies.'[15] They represent ancestral figures or mythical mothers, occasionally forming a 'primordial couple' with a male companion, but 'the power of a man is measured by the number of people under his control, first of all the number of children, which obviously depends on the number of his wives and their fertility. In such societies, then, it is no mystery why women are so often presented in a sexualized way and that this does not contribute to their value.'[16]

In Search of Presumed Religion

Whether the delicate Palaeolithic and Neolithic figures with the volup-
tuous curves were genuinely representations of the mother goddess,
or rather objects of Stone Age pornography, is not only difficult, but
impossible, to ascertain. But perhaps the question is already wrong, and
the answer reveals more about its asker than the object, a product of
misguided expectations.

Perhaps it is wrong to speak of a Palaeolithic religion, since this is to
use a term that is very familiar to us but has no cognate in the world
of that time. For several tens of millennia, small groups had repeatedly
left Africa, travelled shorter or longer distances over several generations
and then either scattered, joined with others or took others in. Thus, the
development of thinking and telling stories about the invisible powers
that determined and influenced the visible world, such as cell growth,
was characterized by repeated isolation and exchange; it is quite plausible
that different groups in Europe, like in Borneo or the Amazon Basin,
formed their own traditions and myths, just as they developed different
languages.

It is impossible to encompass all this with a single term. Burials
and paintings show animist societies and a certain stability of motifs,
techniques and practices, such as the production of those small female
figures; but we can no longer reconstruct what the figures meant to
different people at completely different points in history, what they
thought about them and to what extent these objects were in dialogue
with their intellectual and emotional life. It is easy today to examine
a mummy, trace family relationships through DNA tests or deduce a
prehistoric person's diet from some dental enamel – but knowledge
and memories disappear. No mummy preserves dried-up thoughts and
feelings in its empty cranium.

The diet of prehistoric peoples shows how much information has been
lost. One can establish relatively well from isotopes in dental enamel,

discarded bones and occasionally also plant pollen what basic foodstuffs a community consumed, but not how the food shaped the identity of that community. Eating is never simply eating; it is an instinctive act that immediately turns us into cultural creatures. From the perspective of any food culture, all others seem not only unfamiliar, but also strange, incomprehensible, at times even disgusting. Hunger too is controlled by prejudices, so the hunger of the cave-dwellers will also have had its taboos, its delicacies, its medicines and aphrodisiacs, which reveal the true portrait of a society: the diverse forms of expressing a universal basic need that are strictly codified in themselves, yet perceived as completely natural.

Under this first impression, we find an even more complex web of cultural images that flow directly into the collective myths: what meat is allowed to be eaten? Which animals are legitimate sources of nutrition and which are not? How important is the generosity of hosts? What relationship do people have with hierarchy and surplus, with hunger? How important are ritual purity, hospitality, social status and ritual acts?

From any food culture, one can derive a chain of consequences that intervene in natural contexts: a social claim, a taste, a particular kind of agriculture or a nomadic way of life that follows its food sources, a trading network with access to exotic delicacies, typical diseases – and finally an exhausted biosphere, exterminated animal species and the targeted cultivation of particular plants and animals, invasive species and biotransfer. The cultural construction of food pervades everything that humans touch. Even the Palaeolithic peoples were not exempt from this: they forced back large animals such as mammoths, aurochs and predators and disseminated seeds and pollen on their journeys, thus thoroughly altering the original biodiversity.

Based on the latest research, we can think of our ancestors as Africans. Their skin was probably dark and their hair densely curled, like that of the Venus of Willendorf (if that is indeed her hair). The light pigmentation, better adapted to life in a low-sunlight region like Northern Europe, presumably spread among early humans through relatively frequent relations with Neanderthals. Even today, the genetic make-up of Europeans contains several percent of Neanderthal genes, just as Asian and indigenous American populations carry a larger proportion of genetic material from the likewise extinct Denisovans. The Neanderthals themselves were obviously not simply more primitive lovers or partners of the early *Homo sapiens*; according to the latest research, they also produced cave paintings, bone flutes, stone tools and sewing needles, as well as burying their dead and practising rituals. They were not a precursor to modern humans – they were humans of a different kind.

Since Rousseau, popular projections of a bourgeois culture have included the notion that people who lived 'closer to nature' – whether in the Stone Age or in oral cultures that were only brought into contact with the blessings of the West by colonialists – displayed greater wisdom, modesty and moderation. The environmental historian Daniel R. Headrick warns against idealizing indigenous ways of life; humans have always reshaped their environment in ways that corresponded to their immediate interests.

As early as the Palaeolithic in Central Europe and southern France, and in Colorado and Wyoming long before the arrival of Europeans, entire herds of aurochs, mammoths or bison were driven off cliffs, where most of the animals perished while the hunters only took the prime cuts. In New Zealand, the Māori wiped out the huge, flightless moa bird in search of tender meat to satisfy a busy local market. There is no need to continue this list, but it seems as if the apparent harmony with nature in which many oral cultures were found did not always stem from any deep understanding, but sometimes also from a lack of technological scope. Before the flood too, humans were greedy, hunters tried to hunt more than they could eat and use, and entire landscapes were altered through slash-and-burn clearance.

The people 'before the flood' took their thoughts with them. Their reflexes, and perhaps also a few of their ideas about the world, live on in the people of today along with their DNA; unlike the genetic material, however, they cannot be read and sequenced. Yet there is good reason to believe that they did not correspond to the ideas of nineteenth-century European scientists and museum educators, and were actually surprisingly close to the people of our time, and no less intelligent or creatively talented.

The Dancing God

Art was never bad. It did not only begin after a few tens of thousands of years of helpless scribbling in which there were no recognizable forms. The earliest creative utterances of humans are masterpieces that work so skilfully with the movements and presence of individual animals and entire herds, different moods and colours, that their creators would have been considered great masters in any historical period and any cultural idiom. People with such creative intelligence, so much knowledge about their environment, which they pondered in complex ways and were able to depict as a spiritual reality with such impressive realism, were no fools in other areas either. The horizon of their knowledge was limited, but not much more than was the case for most people in Europe until the early Modern Age. They are known to have traded in luxury goods such as amber and shells from the Baltic to the Mediterranean; as well as the artefacts, they exchanged creative ideas and techniques and, on long evenings, also stories and genetic material.

Not all of our Stone Age ancestors lived in isolation and ate raw meat. Some already lived, long before the introduction of agriculture, in settlements that were perhaps only inhabited for part of the year; they practised division of labour and artisanal mastery, traded, went hunting together, ate and drank together, and may also have made sacrifices as a group – but here the ground of assured facts already becomes shaky.

What did people living in Eurasia 20,000 years ago think about the status of women and the subjugation of nature – and hence, implicitly, that of women by men? It is impossible to reconstruct the world of thoughts before the flood; we do not know whether subjugation already began at that time, as a dream in a cave; as rape, while the twitching firelight illuminated the outlines of giant painted animals on the rock; or as a half-awake idea beneath the endlessness of the starry sky. It would be surprising, however. Other oral societies in Africa and Oceania that have been studied have no concept of humans ruling over nature, and do not tell stories about it, though they certainly know sexual violence. Perhaps the idea of domination only comes about in conjunction with domination of other human beings. The masterful objects of these societies point to a complex thinking that is lost forever, on the one hand, but still remains strangely present, on the other. A figure dances and flickers from the depths of time to the present, as an irritation, a constant question, a denial of order and an affirmation of danger, of the eternal moment, of deadly life. It is the horned god, and one can detect a certain biographical hiccup when he was demoted from the goddess's consort to a mere troublemaker. The divine bulls of the Bronze Age were replaced by religions in which oxen drew carts and bulls were sacrificed to the gods.

The great goddess's lover had already become jobless in classical and classically patriarchal antiquity, for the pantheon was now dominated purely by males. But the horned god was too strong a presence in oral tradition to be eliminated. He conserved his erotic energy as Pan or Priapus, his dangerous aspects as a satyr or the Minotaur, and the leftovers of his creative and destructive majesty as Dionysus, the eternally homeless, horned, drunken god whose ecstatic life force destroys as much as it creates, and who must be tamed, bound, killed and reborn, like the vine itself.

Early Christianity also proved unable to rid itself of Dionysus. More than a few sculptors in the eastern Mediterranean working for customers of different religions made a mistake, depicting the god of Christians with his disciples, his grapevines and his tragic fate as a Greek god, fully naked. But the energy of this ancient god could also be redirected, as shown by the confused sculptures of the messiah in Cairo, Damascus, Jerusalem and Rome. The life of Yeshua ben Yosef from Nazareth was combined with the story of the wandering god and his disciples that had been told for a long time. To neutralize an excess of energy, other aspects were split off into other figures.

The devil, Lucifer, Beelzebub, Krampus, the Green Man, Alpine carnival customs and bull-fighting may have combined old traditions with a new, acceptable interpretation and changed into new-old customs and projections that ate far into the universe of gaming, of alternative spirituality enthusiasts and Wicca, the theology of countless sects and the iconography of evil. We cannot rid ourselves of the horned god, but he becomes more boring in our hands, because his attributes become increasingly similar to one another. But whatever can be interesting, even important, about such uncontrollable forces requires customization.

It remains dark, the world 'before the flood' – even if it is still deep in *Homo sapiens*'s bones. There is no evidence of the delusion of subjugating nature in any archaeological finds or ethnographic comparisons from that distant time. Even if it existed, there is no cave painting or other artefact indicating that humans back then saw themselves as lords of the creation who were superior to nature. Nor is there any nomadic culture today which does not grasp that it is dependent on its environment and its vitality.

Gilgamesh had no access to the knowledge from before the flood, and did not know how to live. So he tried to force not only nature, but even death itself, to its knees. But the people who told one another his story 5,000 years ago were no longer nomads; they lived in cities and villages, they worked in the fields, in workshops, as salespeople or as servants. Although there were still peoples in the Arabian Peninsula, the Caucasus, present-day Afghanistan and Anatolia who had retained their nomadic way of life, these people lived in a new, previously unknown way, and with this life came a new view of the world with a new direction: from top to bottom.

King of the World, King of Assyria

5 The Dying Lioness, fresco from the palace of King Ashurbanipal (669–631 BCE), Nineveh, Mesopotamia. Alabaster, height 160 cm. British Museum, London, Joseph Martin Collection. © The Trustees of the British Museum

The lioness roars one last time, but she knows she has lost this battle. Three arrows have impaled her muscular body, one of them – the sculptor was an exact observer – evidently shattered her lower spine, and she drags her motionless hind legs behind her while looking her slayer in the eye and baring her teeth. This is the moment of her death (figure 5).

The dying lioness is only one detail in a much larger scene that shows King Ashurbanipal of Assyria (who ruled from 669 to 631 BCE) on a lion hunt. Maybe his ancestors set out for the mountains, but in the thoroughly organized kingdom of the most powerful man in the world,

40

lions were caught or bred and then released in an arena so that the ruler could shoot them with a bow from a safe distance, or use a spear from close proximity, always surrounded by a forest of spears and a wall of shields thanks to the royal guard. On one panel in his palace, he is even strangling a lion with his bare hands.

The existence of the palace rests on the logic of killing and being killed: there is only room for one master, the most fearsome predator in this realm. The king of the animals must not simply give way to the king of humans and be annihilated by him – time and again, he is defeated before an audience and spectacularly finished off. This is the mechanism of power.

On the one hand, the palace of Ashurbanipal is a hymn to violence. Its head-high frescoes show the king hunting, waging campaigns, sieges and battles – always victorious, always radiant, always merciless. The bodies of his slain enemies float in the river, are beheaded and mutilated; one prisoner has his tongue ripped out by torturers, two others are pinned to the ground with stakes and skinned alive. This ruler knew no pity and wanted the world to know. Another panel shows him at a banquet, shaded by grapevines, slaves fanning him with cool air while he reclines on his splendid divan and drinks from a bowl.

For all his cruelty as a commander, Ashurbanipal – or Aššur-bāni-apli, as the correct transliteration goes – was anything but a barbarian. His administration was more efficient than anything the world had seen until then. This ruler could not only read and write, but sent his agents to every corner of his kingdom to acquire books, which meant clay tablets inscribed with cuneiform script, for storage in his library. He brought over 30,000 clay tablets to Nineveh, probably the largest collection of texts to date, and a difficult undertaking, since the script itself and some of the texts were already over two millennia old by that time. The king was simultaneously a scholar, an administrator, a military leader and a ruler.

We have taken a huge leap from Uruk around 3,000 BCE to Nineveh in the seventh century BCE – more time than separates the present day from the Qin Dynasty in China or the Roman Republic. Ashurbanipal was an Assyrian and came from the north of Mesopotamia. The bloom of the south and its mighty cities was centuries in the past. And yet there were strong continuities: the same gods were worshipped, the same stories were told. The Sumerian language was still taught and used for ritual texts, and Ashurbanipal was proud of his library, which contained a copy of the Sumerian *Epic of Gilgamesh* that was already centuries old when he acquired it.

Ashurbanipal's palace was a perfect expression of a worldview that had long since internalized the fact that the world is divided between the rulers and the ruled, and that total defeat could only be staved off by demonstrating victory with a strong hand time after time. His palace was a storyboard of his incredible career, from a young man of uncertain origins to the most powerful ruler and fearsome commander of his period, a man whose kingdom extended from North Africa to Afghanistan and from the Mediterranean to the Persian Gulf.

When the King of Babylonia, his own brother, rebelled against Ashurbanipal and declared war on him, he showed no mercy. The betrayed ruler consigned him to the 'burning flames of a conflagration and destroyed him'.[17] He made equally short work of his brother's supporters, as he boasted:

> The chariots, coaches, palanquins, his concubines, the goods of his palace, they brought before me. As for those men (and) their vulgar mouths, who uttered vulgarity against Assur, my god, and plotted evil against me, the prince who fears him, I slit their mouths and brought them low. The rest of the people, [...] at that time, I cut down those people there [...]. Their dismembered bodies I fed to the dogs, swine, wolves, and eagles, to the birds of heaven and the fish of the deep.[18]

By the light of oil lamps or the angled beams of the setting sun, the countless figures on the walls seem to come to life. The quality of these reliefs is still breathtaking today, after almost three millennia. In the place for which they were created and painted with vibrant colours, they would have presented an overwhelming sight to visitors who had travelled for days through the greyish-white barrenness of the surrounding landscape and finally reached the radiant white palace complex. The message was clear and simple: resistance is futile.

This was a Mesopotamian ruler who left behind all boundaries, rhetorically too, precisely because he could pride himself on more than just his bloody victories. The greatest victor of all is the spirit:

I, Ashurbanipal, king of the universe, on whom the gods have bestowed intelligence, who has acquired penetrating acumen for the most recondite details of scholarly erudition (none of my predecessors having any comprehension of such matters), I have placed these tablets for the future in the library at Nineveh for my life and for the well-being of my soul, to sustain the foundations of my royal name.[19]

... and Subdue It

They had miscalculated in a political game with powerful – far too powerful – opponents. Now, from their exile, they looked back at the events of the previous years and regretted the arrogance and depravity for which God had punished them. Judea had never been more than a dusty vassal state, a provincial empire with few subjects, a land of shepherds and farmers with no real power or significance. Travellers knew Jerusalem as a stop on the way from Babylonia to Egypt, the great regional powers of the time.

The elites of the little kingdom were faced with the same problems as any small power with mightier neighbours. The land belonged to the realm of King Nebuchadnezzar (more correctly: Nabū-kudurrī-uṣur II) and had to pay tribute to Babylonia. But when Nebuchadnezzar tried to conquer Egypt in 601 BCE, his army was decimated. The Judeans and a few other tributary provinces seized the opportunity to cast off the Babylonian yoke – or at least exchange it for a more comfortable one: from now on they would pay tribute to Egypt.

The Babylonian king had lost a campaign, but his power within the realm was still intact, and a ruler like him was accustomed to pacifying rebellious provinces and maintaining the flow of tributes and taxes. He marched into Judea, laying waste to the cities on his way, and then besieged Jerusalem, the capital of the renegades, in 589 BCE. After a two-year siege, the city fell. King Zedekiah tried to flee, but he was captured, forced to watch the execution of his son, then blinded and finally sent to Babylon as a prisoner.

Jerusalem fared no better than its king. The Babylonian troops looted everything of value, destroyed the temple and razed the city to the ground. The Judean elites were also forced into Babylonian exile. This act of destruction and banishment was a central trauma for a people that lived in the consciousness of having a special covenant with God, who had promised them prosperity and power in their own country if they observed his commandments. Now he had scattered them, allowed his temple to be destroyed and let his children experience the bitterness of expulsion.

The traumatic experience of the Judeans was simply standard practice in the Neo-Babylonian Empire. Regional revolts had always been the Achilles heel of a great empire, for they not only threatened the ruler militarily and questioned his authority, but also incurred damaging losses of tributes and taxes for the ever-hungry state coffers. One of the most effective methods of pacifying a rebel province was to forcibly resettle the local elites, or even the entire population. Far from their home, their tribes and their power, they no longer posed a threat, while the farmers left behind without political leadership were likewise rendered harmless.

This was also the fate of the Judean elites, probably around 10,000 people, who were resettled from their homeland to Babylonia. Many of the exiles despaired. The famous Psalm 137 shows their emotional fluctuations between sorrow and vengeful bloodlust:

> If I forget you, O Jerusalem, may my right hand forget its skill.
> May my tongue cling to the roof of my mouth if I do not remember you,
>   if I do not consider Jerusalem my highest joy.
> Remember, O LORD, what the Edomites did on the day Jerusalem fell.
>   'Tear it down,' they cried, 'tear it down to its foundations!'
> O Daughter of Babylon, doomed to destruction, happy is he who repays
>   you for what you have done to us – he who seizes your infants and
>   dashes them against the rocks.

This Babylonian captivity was probably not as terrible as the Bible, for more or less propagandistic reasons, makes out. We know from historical documents that Judeans worked as officials in the Babylonian administration and were evidently integrated into society. The religious trauma, however, was all the deeper: God, who had made a covenant with them and led them into the land in which milk and honey flow, had cast them aside. But what is a people worth after its god abandons it? Had they all sinned, or was their god not the great creator after all?

The Babylonian exile was a time of theological and existential crisis for the Judeans. They could not sing any songs to their lord in this foreign land, but they could reshape their relationship with him and at least build a new country, a new homeland, until their return – in a book. This is where the human self-image that Bruno Latour calls the metaphysical legacy of the Galilee, and Heinrich Heine the Nazarene spirit, came into being. Here subjugation was turned into a scripturally fixed guiding principle and divine mandate.

In Babylonian exile, the Judeans became Jews – not only because the experience of exile was so inextricably bound up with Jewish history, but also because the banished elites set about gathering the traditions of their people handed down in writing and by word of mouth, making a selection and determining a canonical version.

The Babylonian exile is the period in which the five books of Moses, constituting the Jewish Torah and the centrepiece of the Bible, were set down by scholars in their final form. A central part of the Bible as we know it today came about there. The Jewish community of Babylonia clung to these Holy Scriptures and devotedly began to interpret the word of God, whose content it had only just decided on in a kind of committee, down to the last dot on the last letter. The people of the book were born.

In this situation of exile and religious crisis, the most important thing was for the new Holy Scriptures to have the necessary legitimacy to be accepted by everyone, for the editing did not take place in a vacuum, but within the life of a traumatized community comprising different political and religious factions.

Presumably the scholars had brought their own Holy Scriptures from Jerusalem, as well as passing on other episodes, psalms, lists or passages orally, with different sub-communities considering different traditions sacred. All these interests had to be taken into account when finalizing the text. It is likely that several alternative versions of the holy texts were circulating among the exiles, which would explain why two creation stories found their way into the final version, not to mention numerous other inconsistencies. The scholars who had known these texts by heart since childhood were aware of the contradictions, but the Holy Scriptures would not have been accepted by all groups involved if they had not recognized their own traditions within them. This resulted in a document that frequently contradicts itself – but it also opened up almost infinite room for interpretation.

In addition to their own traditions, however, the editors in Babylonian exile also drew on the myths of their surroundings. Many of these myths were circulating throughout West Asia and the Levant and in all Semitic cultures, from Uruk to Anatolia and from Jerusalem to Egypt, and may also have entered the Bible in other ways; but many of the elements seem to have been taken very specifically from Mesopotamian myths that were already millennia old by then.

It is not only Uta-napishti and the story of the flood that have clear biblical parallels. The creation stories (which were handed down in Mesopotamia over millennia in different versions) are similar in many details, from the creation from chaos to the separation of heaven and earth. The similarities continue from there: the goddess Inanna, who leans against a tree in a garden and accepts a fruit from a serpent that lives there; the origins of the legendary King Sargon of Akkad, who was put in a basket after birth and placed on the river by his mother, just like Moses in Egypt; the Tower of Babel; the many civil laws and penal rules, which precisely correspond to the Code of Hammurabi, specifying the same punishments or compensations for the same offences with the same type of language: stylistic devices like rhetorical repetitions; the verse forms and images used – all this anchors the Bible and its origins firmly in Babylon, even if archaeological finds show that the *Epic of Gilgamesh* was already known in Canaan in 1400 BCE, which means that the Mesopotamian influence had already started before the Judeans arrived in Babylon.

The Babylonian stories are considerably more entertaining than the biblical ones, for their protagonists are gods who are every bit as greedy, stupid and lustful as their creations, humans, and the resulting conflicts and situations always provide material for good – indeed divine – soap operas. The Mesopotamian pantheon (which changed over time, but retained a surprising number of its main figures) is refreshingly oriented towards the all-too-divine needs of its inhabitants. As mentioned above, for example, the first cosmic battle between the gods arises from the father god's wish for quiet. When a younger generation of gods disturbs him with their incessant noise, he summarily decides to murder them all. Another god stands up for them, which immediately results in a war that ends with the Earth being formed from the body of a slain goddess.

Humans only appear later, as a particularly innovative idea from their creators, for until then the gods had to labour in the fields to sustain themselves. They create humans so that they no longer have to do any hard work, and can instead do what Goethe's Prometheus would later accuse the Greek gods of doing: 'Ye nourish painfully, / With sacrifices / And votive prayers, / Your majesty; / Ye would e'en starve, / If children and beggars / Were not trusting fools.'[20]

In Sumerian mythology, the next problem is that even the trusting fools bring intolerable unrest into the life of the gods, which leads the gods to decide on destroying their creation because it does not meet their expectations. And so they send the flood – only to regret it sorely, as the disappearance of humans puts paid to their source of food and they remain hungry until humans can make sacrifices again.

The editors of the Bible gave their god a harsher character, eliminating the more frivolous elements of the Babylonian pantheon. The Lord of the Judeans was a jealous and vengeful god, but he had great plans for humanity. After creating Adam and Eve, he gave them an assignment: 'God blessed them and said to them "Be fruitful and increase in number; fill the earth and subdue it. Rule over the fish of the sea and the birds of the air and over every living creature that moves on the ground."'[21] These lines would have appealed to Gilgamesh, who had set out to prove his dominance and his power, to slay the forest spirit, clear the holy cedar forest, kill the Bull of Heaven and overcome death itself. In imperialist Babylonia, whose king boasted of his conquests and victorious battles, it was hardly an outlandish idea to want to subdue the Earth itself.

The political situation of the Jewish exiles in Babylon was not such that it could offer them any realistic prospects of earthly power. They were prisoners of the state, forced evacuees, lived in a foreign land and were viewed with suspicion. The ambition to subjugate the Earth was a fantasy, a rhetorical position that was perhaps meant to embolden or was formulated out of defiance, but did not correspond to any reality. The Bible became not only the homeland of a people, but also an imaginary place in which they could be someone else: no longer slaves and exiles, but rather masters in their own house.

Lost in Translation?

The formulation of the Bible caused considerable discomfort to later generations of theologians. Was one message of the Bible not that humans should be good shepherds of the creation – empathetic helpers?

The Hebrew word that was placed in the Lord's mouth is וְכִבְשֻׁהָ (*va-kibsu-ha*), literally 'and you shall subdue it'. In Wilhelm Gesenius's *Hebrew and Chaldee Lexicon to the Old Testament Scriptures*, the verb on which the command is based, כָּבַשׁ (*kabash*), has the following meanings: 'To tread with the feet, to trample under feet [...] to subject, to subdue to oneself',[22] with a note that in the book of Esther (7:8) it can also mean 'to force a woman' – that is, to rape. *Kebesh* means 'footstool'. In related languages, from Aramaic and Syriac to Babylonian, the corresponding verbs have a very similar, unmistakably violent meaning.

Gesenius's lexicon was written in the early nineteenth century, but the translation of the phrase *va-kibsu-ha* also has a very clear tradition of translation that determined over generations how this passage was read in Europe by people who were unable to read the Hebrew original – which made these versions the historically influential ones. In the Vulgate, the Latin translation from the fourth century, it reads: *Crescite et multiplicamini, et replete terram, et subjicite eam*. Here *et subjicite eam* is translated as 'and subdue it', and would normally refer to overpowering or subjugating an adversary or an animal – that is, an antagonistic but inferior will. For Catholics, this would henceforth be the dominant reading of the text, albeit not the only valid one.

Other Christian confessions and languages took the same line, however. In the King James Bible published in 1611, all English-speaking believers read: 'Be fruitful, and multiply, and replenish the earth, and subdue it.' In his translation from almost seventy years earlier (1543), Martin Luther makes an addition after 'subdue it': 'What you build and work on the land shall be your own, and the Earth shall serve, carry and give to you for this.'

Further biblical passages also refer to the exceptional position of humans in comparison to the rest of creation. In Psalm 8, for example, the psalmist asks:

> When I consider your heavens, the work of your fingers, the moon and the stars, which you have set in place,
> what is man that you are mindful of him, the son of man that you care for him?
> You made him a little lower than the heavenly beings and crowned him with glory and honour.
> You made him ruler over the works of your hands; you put everything under his feet.[23]

The meaning of the divine commandment thus had a stable tradition, outside Christianity too. The patriarchal interpretation can also look back on a long history. The great biblical commentator known as Rashi (whose full name was Rabbi Shlomo Yitzchaki, or, in French, Solomon de Troyes, 1040–1105) is known for his gentle, rational interpretations of the Holy Scriptures. Following the rabbinical predilection for inter-pretive language games, he suggested an alternative reading of the phrase 'and you shall subdue it', taking its meaning as 'and you shall subdue her', which teaches the reader 'that the male controls the female in order that she may not become a gad-about'.[24] This is an astonishing shift of meaning: the one who is tasked with domination is unambiguously male. The woman was not herself a subjugator, but had to accept the role of the subjugated.

Look on My Works!

The biblical command to subdue the Earth has had a huge career. Before we retrace this development, however, it is interesting to pause briefly at the moment of its birth and explore what this meant in the context of its time.

Gilgamesh was doomed to failure because he was too high-handed, too self-aggrandizing, because he did not know the customs from the time 'before the flood' and antagonized the gods. At the end of the story, having failed, he is wiser. But which laws had he broken? In a polytheistic world where gods, demons, spirits, nymphs, ancestors and harpies or their respective equivalents are at work (whether ancient Greece, Korea or Tahiti), it was clear to everyone that anything they began could only succeed if they came to an arrangement with the appropriate deity. These arrangements could take different forms, but they usually involved sacrifices; in the distant past, human sacrifices may have played a central part. In Homer's *Odyssey*, Agamemnon even had to sacrifice his own daughter to ensure a favourable outcome for his campaign. Christianity is divided into theological schools: those that intend Holy Communion only to commemorate the sacrificial death of Christ, and those that believe that the wafer and wine genuinely become the body and blood of the Lord.

This sacrificial practice grew from the knowledge that every change in the physical world affects the interests of something else – a being that symbolizes natural forces or is identical with them, another will or another power. These natural forces can be useful or harmful to others, depending on which god or which gods are more powerful, which sacrifice is generous enough, which request sufficiently heartfelt, which other beings are a goddess's enemies, whom a god wants to entice into his heavenly bed, what gods and humans desire and how these different interests can be balanced with one another.

Life in a polytheistic world is a constant give and take. Everything I do comes into contact with an invisible but very lively, very real power that has its own interests and whose help I will need. It would be foolish to offend the forest spirits before setting off on a journey, otherwise I will lose my way or find myself in a storm. I cannot kill an animal without making an offering, or fell a tree without spreading seeds, for I am not alone; my life does not proceed in isolation, for every successful action comes from a mutual recognition of all parties involved, their active assistance and the appropriate compensation. What goes around comes around.

This polytheistic meaning of life was so widespread that we only know of a few exceptions, whose worldview had similarly animistic origins and presumed spiritual actors to be behind natural phenomena, but saw themselves as opponents of that world and believed they were successful in outsmarting the invisible powers. As far as we can tell, however, no known culture assumed that the material world was dead, that humans existed as the only higher life form in the world and that they could act without consulting the invisible powers and making allies of them, inevitably becoming entangled with the visible and invisible life around them.

Against this background, the biblical idea of dominating nature resembled a mythological nuclear bomb. Instead of representing the natural world as ensouled and full of actors with whom one had to reach an understanding, the Bible of the one god knew only a dead Earth, an inanimate world of dust with no will or power of its own, merely waiting to be subjugated, ploughed, possessed, sold, penetrated and fertilized, bought and sold. This upheaval of perspective has the potential for enormous self-empowerment. Man is no longer a slave to the gods – he is the lord of creation.

This explosive idea that God created humans to subjugate and dominate the rest of creation, and that this creation no longer exists in a reciprocal relationship with its human inhabitants, but is completely passive and enslaved, was taken up into the Hebrew Bible, even though its authors were devoid of any political power – or perhaps for that very reason.

After the Jews returned from Babylonian exile in 539 BCE, this situation initially remained much the same. The eastern Mediterranean and the Levant were ruled by changing regional powers. The Babylonians and Egyptians were finally followed by the Romans, who made Judea their province: a restless place, not especially lucrative for its governors but always bubbling with rebellious energies; it was not a popular post in the Roman Empire. The fate of the Jewish elites depended on their alliances; the colonial yoke never rested lightly on their shoulders, and finally a new rebellion broke out, with a gruesome, albeit predictable, end: in 70 CE the Romans, as a great power that had been vexed, destroyed the second temple too – the beginning of the Jewish diaspora.

The explosive idea of dominating nature went almost unnoticed in the Holy Scriptures of the Jews for six centuries. It could have stayed there and died slowly, one of the many redundant theological concepts, unexplained fragments and mythological leftovers that one finds here and there in the Bible.

Elsewhere, this idea had died long ago. King Cyrus (c.600–530 BCE), who had allowed the Jews to return to their land, was no longer a Neo-Babylonian monarch, but rather a Persian who had invaded Babylonia with a huge army and captured Babylon itself. He proclaimed himself the legitimate successor to the kings who had ruled there, but his reign marked the beginning of the end of the realm, the direct successor to a tradition spanning five millennia. The power of Mesopotamian societies had finally been overcome, and the new masters were gradually accompanied by new stories and myths, laws and customs.

This transition from one world into another was hotly contested, however. The Achaemenid Cyrus and his successors were not recognized by the Babylonian elites. Time and again, there were uprisings, rivals staked claims to the throne, and civil war spread and smouldered over centuries with repeated outbreaks until King Xerxes finally destroyed Babylon in a punitive expedition and razed the city walls to the ground. Babylonian culture survived into the Hellenistic period, but these were the last remnants of a dying civilization.

Nineveh, the splendid capital of King Ashurbanipal, lord of Assyria and lord of the world, owner of the first systematically ordered library in history, victor over countless foes and much-extolled triumphator over enemies and lions, which he strangled with his bare hands – the glorious Nineveh lay in ashes. Buried beneath its ruins lay the immense library that would come to light more than two millennia later.

In 1807, when the fascination of the imperial British for further world empires and their historical fate reached its climax, the English Romantic Percy Bysshe Shelley wrote a poem entitled 'Ozymandias', another name for Pharaoh Ramses II. He met a traveller, the author reflects, who told him of two 'vast and trunkless legs of stone' standing lost in the desert, and next to it a 'shattered visage' with the 'sneer of cold command' on its lips. There was an inscription on the ruined statue's pedestal that fascinated the poet:

> My name is Ozymandias, King of Kings;
> Look on my Works, ye Mighty, and despair!
> Nothing beside remains. Round the decay
> Of that colossal Wreck, boundless and bare
> The lone and level sands stretch far away.[25]

The end of the great Mesopotamian empires and their city states was not as sudden or dramatic as Shelley suggests with his formulation, which may have been a warning to his contemporaries (his wife, Mary Shelley, wrote a further warning with her novel about Dr Frankenstein). Even the greatest empires come to an end, and even near-absolute power may leave little more than the dusty ruins of its ambitions.

The idea of dominating nature would probably have died a slow death along with the Neo-Babylonian Empire as part of a vanquished and powerless culture, but it had already spread, infecting a wider culture and even developing beyond the ultimately naive Mesopotamian concept of power. It may have impressed a royal court when Ashurbanipal, standing in his chariot in the arena, surrounded by shields and spears, fired arrows at a lion that had been caught – or even bred – for this, perhaps, or even choked it to death with his bare hands once it was weakened and surrounded by spearheads, glinting in the sun, ready to affirm the king's godlike authority at any moment with a spirited killing thrust. But lord of the whole creation? Over all animals? Over everything that grows? That was a different kind of power.

This inflated absolute power remained trapped in a few letters, with no correlate in reality. It did not even appear in rituals. It remained a dream that never really managed to occupy centre stage, dreamt by a rebellious people in an unloved province of the Roman Empire, and soon also by a mysterious Jewish sect that would gather ever more followers. Its founder was a charismatic preacher named Yeshua ben Yosef whose words fitted perfectly into a time of apocalyptic expectations, of rebellions and wandering preachers, who belonged to the scenery of ancient cities and drew especially large crowds in the volatile atmosphere of Judea.

The monotheistic idea had a hidden impact in the ancient world, where a seemingly endless carnival of competing, exotic Olympians were honoured alongside one another in a kind of religious hypercapitalism. In the middle of this bickering between Jupiter and Mithras or Isis and Astarte, it would have been a rhetorical bombshell for the empire's philosophical minds to install a god who stood above this undignified wrangling: a single, universal, abstract, all-powerful, all-knowing god, a god who would have met with the great Plato's approval. But the god of the Jews was clearly ineligible, since he had tied himself exclusively to this small desert people. Neither his commandments nor his admittedly shaky loyalty applied to gentiles, and whoever wished to follow him had to accept such rituals as circumcision and abstain from pork, which Romans would have found odd at best. No, it was not the god of the Jews.

However, the Roman elites were also repulsed by the spectacular failure of this impotent god, who had allowed himself to be slaughtered like a donkey – which is why, in a famous graffito from the Roman catacombs, he was depicted as a crucified man with a donkey's head. Rabbi Yeshua had been accused of rebellion and crucified like Spartacus, leader of the insurgent slaves of Rome. For reasons that were incomprehensible to the power-loving Romans, however, it was precisely that shameful end and the failure of his inflammatory speeches that had created a movement which not only outlived him, but was also opened by one of his disciples to non-Jewish recruits who wanted to save their souls, but considered the observance of Jewish laws – and especially circumcision – too high a price to pay. In its new form, the community grew all over the Roman Empire. The anarchistic message of poverty, solidarity and transformative suffering had a magnetic persuasiveness amid the cynical corruption of the Roman rulers and their purely business relationships with their countless gods.

And then the former Christ-hater and lifelong woman-hater Paul came on the scene, having fallen off his horse on the proverbial road to Damascus and been converted by a vision. He gave access to a universal god by claiming that the covenant with him had been renewed, and declared the rebel preacher Yeshua ben Yosef the anointed, the *Christos*, the saviour of all humanity.

For the intellectuals of the ancient world, who were tired of Aristotelian technical exercises and corruption dressed up as religion, this excitingly anarchistic message offered an immensely interesting possibility to rethink the world. A patchwork world of gods with contradictory legends and traditions and rival deities did not fulfil the need for order and transcendence that had already moved Plato, and which the Neoplatonic philosophers had brought up for discussion once again. Did an empire with a Caesar not need a god as well? And should this god not be cleansed of all contingencies and imperfections? And would this god not give all people in his kingdom civil rights, as Rome did with its inhabitants?

The Triumph of Light over Darkness

No one captured the excitement over the new religion in as lively a way as a North African bishop born in the small coastal town of Hippo, in present-day Algeria: a certain Augustine (354–430): 'All are astonished to see the entire human race converging on the Crucified One, from emperors down to beggars in their rags.'[26]

The new religion released a social and philosophical explosiveness that evidently made it irresistible. Augustine himself had only discovered this exciting doctrine as an adult; he came from a wealthy family in North Africa and enjoyed an excellent education. He had a mistress who bore him a son, then travelled to Rome and Milan to continue his studies and teach rhetoric himself. He had a brilliant mind and lived life to the full, but he was also a seeker who experimented with a variety of philosophical schools, cults and sects, yet had not found a spiritual home anywhere.

After his conversion to Christianity, he gradually left behind his luxurious way of life. The former teacher of eloquence became one of the greatest preachers and theologians of the ancient world, a personality with amazingly modern aspects. His work *Confessions* was not only a monument to the struggle in late antiquity with personal convictions and inclinations; it also founded the genre of autobiography and systematic self-reflection.

As a young man, Augustine had been especially taken with the Manicheans and Neoplatonists, and one still finds traces of both intellectual worlds in his theological writings. The Manichean doctrine held that good and evil exist and that, as well as the good god, there is also an evil one who seeks to undo all his works. Earthly existence was a battle between the sons of light and the sons of darkness.

The Neoplatonists had a more subtle intellectual edifice, as befitted what was a cutting-edge school of philosophy for a cultivated, urbane elite. Their thinking sprang from Plato's idealism, especially the idea that genuine truth is to be found in the pure forms and concepts of the spirit, and that the material world is no more than a representation of this ideal world, as he showed in his famous cave allegory. People sitting in a cave with their backs to the entrance, unable to move their heads, must take the shadows on the back of the cave for reality, even if they are only two-dimensional reproductions of the genuine, real beings standing at the cave entrance.

It followed from this that the best way to grasp the truth about the world was through the contemplation of pure ideas or forms, not empirical research. The earthly world, after all, was no more than the base, impure reflection of the ideal world, and the goal of a philosopher's life had to be to shed these constraints of earthly life as far as possible to free one's own thinking, ideal self – which could partake of the world of pure forms – from the clutches of the material world.

At the height of the Roman Empire, it was Plotinus, an Egyptian thinker who lived in Rome, who founded a school of thought which revived the thought of Plato – an aristocratic philosophy that gave ambitious minds what the polyglot divine circus of the official religions could not offer: order, clarity, principles, orientation. The young Augustine likewise tried to create a harmonic totality in his spirit from the concert of different traditions, religions and explanations of the world with which he, as the son of a Christian mother from the Berber people and a heathen father, had grown up in North Africa.

For his biblical interpretations and sermons, Augustine's experience as a rhetorician and student of various philosophical tendencies came in handy. In a passage on the special position of humans in relation to nature, Plato's influence is clear:

> Whereas, then, the omnipotent God, who is also good and just and merciful, who made all things [...], made also man after His own image, in order that, as He Himself, in virtue of His omnipotence, presides over universal creation, so man, in virtue of that intelligence of his by which he comes to know even his Creator and worships Him, might preside over all the living creatures of earth.[27]

This argument encourages the emerging delusion of dominating nature. In the creation story, it was not reason that set Adam and Eve apart from all other creatures. There is no reference to the reason or intellect of God's creatures; God created them in his image, but this is a visual metaphor. Of course, like other parts of the Bible, the book of Genesis was passed through the filter of Neoplatonic reason and systematics to transform the muddled fantasies of a completely different tradition into a homogeneous dogma.

It is the form of reason cleansed of all passions that enables humans to become almost divine and partake in the essence of God. The church father did not fail to offer proof of this:

> For all the other animals are subject to man, not by reason of the body, but by reason of the intellect which we have and they do not have. Even our body has been made so that it reveals that we are better than the beasts and, for that reason, like God. For the bodies of all the animals which live either in the waters or on the earth, or which fly in the air, are turned toward the earth and are not erect as is the body of man. This signifies that our mind ought to be raised up toward those things above it, that is, to eternal spiritual things. It is especially by reason of the mind that we understand that man was made to the image and likeness of God, as even the erect form of the body testifies.[28]

This was a classic conclusion by analogy that should really have been beneath such an otherwise impressive thinker. Whatever is turned towards the earth is low and must be dominated, but whatever rises higher is created for a contemplation of the divine. Had Augustine, an African, ever seen a giraffe?

One senses a certain pressure to justify the exalted position of humans. Elsewhere, Augustine asks how it can be that humans were given dominion over animals, yet wild animals can be dangerous and harmful to them and cannot be controlled. Here, too, he digs into his box of rhetorical tricks. If, following the Fall of Man and the loss of paradisaic existence, humans still have power over pets and livestock, one can imagine how much greater this power was before Adam took a bite from the wrong fruit.

To Augustine, there was a fundamental difference between humans and the rest of nature. In one of the more politically controversial parts of his reflections, he concluded that God himself had given his favourite creature unlimited power over nature, but not over other humans. A being endowed with reason must therefore never be enslaved, he argues, and slavery is a moral evil; but then he states that original sin makes slaves of us all, and the existence of slavery is thus acceptable.

Augustine inveighed against the tyrannical rule of people over other people and held that every person has the same dignity, even if this was not always the case in practice. While he was at least ambivalent about slavery, he did, as one of the most important theorists of the just war, support other forms of violence. The idea of a Manichean conflict between good and evil may have faded in Augustine's thinking, but it still formed the framework of his moral ideas. He viewed intra-clerical and political disunity as an affront to the truth, and acknowledged that vehement conflicts were inevitable and he could, on the side of Christ, wage a just war that gave him a robust mandate to use force. There is nothing wrong with war, he writes laconically, for 'people die who will eventually die anyway'. Only a coward complains of this:

> No one must ever question the rightness of a war waged on God's command, since not even that which is undertaken from human greed can cause any real harm either to the incorruptible God or to any of his holy ones. God commands war to drive out, to crush or to subjugate the pride of mortals. Suffering war exercises the patience of his saints, humbles them and helps them to accept his fatherly correction. No one has any power over them unless it is given from above. All power comes from God's command or permission. Thus a just man may rightly fight for the order of civil peace even if he serves under the command of a ruler who is himself irreligious.[29]

With his *bellum iustum*, the church father provided the justification for the Crusades, colonial expeditions and holy wars centuries in advance – even if they would probably still have taken place without him. His holiest Manichean wrath, however, was reserved not for any outside enemy, but sin itself, and thus human nature, which shackled his immortal soul and godlike reason to a dirty, lustful, corrupt body. It is on this stage that the cosmic war of annihilation between good and evil is waged – for sin, that most thankful of protagonists, was not only committed by individuals in moments of carelessness, but rather passed on through the body. Humans are born with original sin, guilty from birth, damned and destined for the depths of hell, where eternal punishment awaits them. Adam's disobedience and Eve's temptation followed humanity as a shadow cast by the fires of hell. Only the undeserved grace of God could save souls from eternal damnation.

'*Caro tua, coniunx tua* – your body is your wife', Saint Augustine preached – on the one hand, demanding that one respect the body, and, on the other hand, clearly thinking of the seductive Eve; for this erstwhile bon viveur had developed into an ascetic theologian who recalled only too well the Manichean teaching that evil comes into the world through sex and poisons the souls of men with lust. Marriage is therefore a *remedium concupiscentiae*, a cure for lust, for complete abstinence is even better than marital sex, even if only few manage to uphold it. Augustine also mentioned that the love of the most important woman in his life, his mother, was in a sense 'too carnal', which has repeatedly led scholars to believe that his relationship with women in general may not have been exactly simple or conflict-free.

At this point, the history of subjugation turns inwards. Uta-napishti had already explained to Gilgamesh that he must first learn to control his own nature before he could attain immortality. Before Augustine, however, the command to subdue the Earth had been understood in the literal sense – namely, as Jewish settlement and military subjugation – or in the sense of creation through the magical naming of all animals. Influence on nature remained small, however, with the rhetorical position in the Bible having no real-world consequences. The Jews and later the Christians were subjected to tempests and earthquakes, as well as the brutality of the Roman army.

Augustine transposed this dynamic inside humans: from now on, the church's doctrine of virtue would always revolve around the control, suppression and sublimation of desire and bodily impulses, which were conceived of as inherently sinful and depraved. The division of the human being into a pure spirit and a body that is also metaphysically dirty was complete.

The Bishop of Hippo even distrusted pleasure, perhaps especially as someone who had described himself as a sensualist, because it could not be consciously controlled. 'What friend of wisdom', he asks, 'would not prefer, if this were possible, to beget children without this lust, so that in this function of begetting offspring the members created for this purpose should not be stimulated by the heat of lust, but should be actuated by his volition, in the same way as his other members serve him for their respective ends?'[30]

The 'function of begetting offspring' and the memory thereof clearly afflicted the holy man. This marked the beginning of a 2,000-year literary and artistic tradition of prudishness, even if there were precursors to whom Augustine himself pays tribute, such as the Roman poet Virgil (70–19 BCE).

Not all Romans were as fun-loving as their art and the history of their rulers suggest. A class of disillusioned aristocrats had turned away from the orgies and intrigues, pretty slave girls and toy boys, bread and circuses. They searched for the truth in the philosophy of the venerable Plato and the battle against sensuality. Virgil described the human body as a terrible burden: 'The life-force of those seeds is fire, their source celestial, / But they are deadened and dimmed by the sinful bodies they live in.'[31] The fiery power from the heavens that dwells within humans must fight heroically against the body and its influences, because: 'Whence these souls of ours feel fear, desire, grief, joy, / But encased in their blind, dark prison discern not the heaven-light above.'[32]

The great orator and sophist Augustine managed to give this joyless conception of humans a despairing twist, for while Virgil merely saw his body as the enemy of his soul, original sin was more deeply buried. The sin lay in Adam's disobedience, in a stirring of the soul: 'it was not the corruptible flesh that made the soul sinful, but the sinful soul that made the flesh corruptible'.[33] The dance between body and soul, sin and salvation, becomes a hopeless chase around an empty centre. But the solution is clear: a life based on the spirit and the battle against fleshly desire; abstinence and a merciless war on sin. There is nothing more life-denying and depressing than reading Augustine's deliberations on the 'shame which attends all sexual intercourse';[34] but, like various other celibate thinkers, he wrote about it in almost obsessive detail.

It was not only Augustine who preached an extreme worldview woven together from contemporary influences and probably also biographical motifs. Christianity was by far not the only religion in Mediterranean late antiquity that experimented with radical ideas and created ever new schisms and sects. It is especially interesting, however, because, unlike other cults from antiquity, it won the battle for the future. Thus, Christianity secured the survival of the West Asian delusion of subjugating nature within a larger context. With the Christianization of the entire Roman Empire, and later of Europe too, an obsession fuelled by Greek systematics, Platonic hostility to pleasure and Manichean paranoia advanced to completely new territories.

The leading thinkers of the new religion saw the conquest of these territories as a holy mission from the start, and the battle raged not only geographically, but also within every society, for the world was divided between those 'who wish to live after the flesh' and those 'who wish to live after the spirit'.[35] History comes about through this battle, as Augustine makes clear in the first words of his magnum opus *The City of God*: 'For to this earthly city belong the enemies against whom I have to defend the city of God.'[36]

'Living after the flesh' and in sin was the consequence of Eve's gullibility and Adam's weakness. Augustine did not ask why it was wrong to eat from the Tree of Knowledge or why God had forbidden the fruit of this tree in particular; nor did he address the many serpent myths of his time, but rather deduced the wretchedness and guilt of all humans from the first two:

> for by them so great a sin was committed, that by it the human nature was altered for the worse, and was transmitted also to their posterity, liable to sin and subject to death. And the kingdom of death so reigned over men, that the deserved penalty of sin would have hurled all headlong even into the second death, of which there is no end, had not the undeserved grace of God saved them therefrom.[37]

In the name of this God, the Good News was brought first to one continent, and finally to all.

It was not only in the works of Augustine that Christianity and Greek thought joyfully embraced. It was said of Origen of Alexandria (185–253 or 254) that his repulsion towards his own body and his despair at his own uncontrollable desire induced him to castrate himself, in order to work on his theological writings without being plagued by temptation.

According to his biographers, he had already internalized the sinfulness of his body as a young man. Like many people who thoroughly question their own world, Origen had a history marked by personal tragedies and conflicts. His father, a devout Christian who refused to sacrifice to the

Roman gods, was publicly executed when his son was barely seventeen years old. Origen had tried to turn himself in to the authorities in order to die a martyr's death too, but legend has it that his mother hid all his clothes, and the shame of his nakedness was more terrible than the missed opportunity to become a martyr.

In more than 2,000 treatises, Origen sought to reinvent Christianity as a morally strict religion of reason, drawing on the Platonic theory of forms. The doctrines and ideas of this Middle Eastern sect first had to be subjected to a rigorous redefinition by the philosophers of antiquity. They often found the words of Jesus simplistic, devoid of any learnedness or rhetorical sophistication: the words of a carpenter, not a cultivated person. As for the rest of the Bible: muddled stories, endless lists, contradictory passages, murky legends or obvious borrowings from other cultures. The raw product was positively crying out to be refined through Plato's theory of forms and the relentless systematics of an Aristotle. Yeshua ben Yosef, who may have been wearing nothing more than a simple woollen tunic for his shameful crucifixion, entered the stage of the debate wrapped in the flowing toga of a classical philosopher.

Was the Emperor Constantine acting out of profound conviction or political calculus when, before a decisive battle in the year 312, he placed himself under the protection of Christ, struck out under the sign of the cross and triumphed? The dispute over this will continue for a long time, but is actually unimportant. The fact that, after winning the battle, he placed the severed head of his foe on a lance for a victory parade perhaps shows that Constantine still had a thing or two to learn about Christian virtues; but this Christ had, after all, just brought him triumph in a gruesome slaughter.

The triumph of Christianity was accompanied by that of the idea of subjugating nature. The Romans had always lived by the sword, but their world domination was limited to the capacities of their armies and communication channels. Now, however, a new form of power emerged, as the historian Tom Holland observes: 'Formed of a great confluence of traditions – Persian and Jewish, Greek and Roman – it has long survived the collapse of the empire from which it first emerged, to become, in the words of one Jewish scholar, "the most powerful of hegemonic cultural systems in the history of the world".'[38]

With the Christianization of the Roman Empire, an *ecclesia* that had emerged from a state-threatening Levantine sect was transformed once and for all into a world power that, from the start, resembled the Roman state more than the dreams of a failed messiah. Christianity learnt to rule, to wage wars, to punish, all in the name of the Lord; for centuries, Christ would be portrayed not as a crucified man but as a radiant king, sitting on his throne against a golden background and surrounded by his court. In a mosaic in the church of San Vitale in Ravenna (see figure 6), consecrated in 547 CE, Christ sits enthroned as a beardless Caesar with a gold-edged toga on a blue sphere symbolizing the universe.

The four figures around him are also wearing Roman dress. In their togas, the two angels look like winged administrators of their heavenly emperor. Their feet are on the earth, a thin strip of green with a few flowers on it; but their bodies stand out against a golden background, golden like Christ's halo, sharing in his glory through their mere physical elevation over nature, just as Augustine had argued. The *pantocrator*, the ruler of all depicted by the early church, gives the saint the martyr's crown and the bishop a church; he distributes power among his own so

6 Christ as *cosmocrator*, 'ruler of the world', enthroned against a golden background and surrounded by his court. Christ hands San Vitale the martyr's crown; an angel gives Bishop Ecclesius a model of the church. Mosaic in the Church of San Vitale in Ravenna, consecrated in 547 CE

that they will go out and spread his word. If the legacy of ancient Greece had lent the young religion intellectual respectability and a wealth of long-structured ideas, it was Rome that gave it muscle.

The rest of the history of European Christianization is well known, and although the theological debates, disputes and campaigns did not settle down for centuries, the framework for the church's conception of humans and the relationship with nature had been firmly set. On the territory of theology, several written traditions came together that were each accompanied by entire clouds of debates, myths, standpoints, strategies, memories and arguments that could be written, copied and sent to every last corner of the empire unchanged. An Irish monk could read the words of the North African Augustine, and both were part of a tradition that extended from Mesopotamia and even further back, into the mists of the first civilizations.

This shared tradition owed its intellectual and cultural richness to the fact that it absorbed a variety of others and used them to create a new, often bizarre constellation. The message of the Gospels and the brutal power politics of the Roman Empire had long been diametrically opposed to each other. Christians had been persecuted in ancient Rome because they refused to participate in public sacrificial rites. The Romans were surprised by the naive idea that one needed to believe in the literal truth of a myth to partake in the appropriate ritual. They found this intellectually backward, but were most bothered by the lack of social solidarity. Whoever did not take part in the rituals was not showing themselves as a Roman patriot, whatever Lares or philosophers they honoured at home. And if someone also belonged to a sect that emerged from a rebellious people and worshipped a god who stood not for power but for powerlessness – anyone who brought together so much perverseness could only expect to be thrown to the lions as an enemy of the state, to the enjoyment of the paying audience beneath the awnings of the Circus Maximus.

Whether spiritual event or political trick, Constantine's conversion changed the course of history. It added a further contradiction to the explosive ideas of what was by now a complex and contradictory intellectual tradition from Athens and Jerusalem, from Mesopotamia and North Africa, and simultaneously equipped it with a network of power.

For Constantine, the one god was a clearly unique feature that also allowed him to attack the jungle of cults from every corner of the empire – which, not by chance, were often connected to the political interests of his rivals. The empire had a new image: a heavenly emperor whose radiance also emanated from his earthly governor. This practical message was accompanied by the entire administrative network of the empire, the military power of his legions, the streets and ports and libraries and schools of rhetoric, the provincial elites and their love for Roman fashions, as well as the dynastic power games, where conversions play a strategic part and new allies can be won. The levers of power were immensely effective.

The profound contradictions and translation problems between the Bible and classical thought (to say nothing of the other influences) had attracted the best talents of theologians, philosophers and other shrewd minds. Was Plato's pure form really the god of the Abrahams and Isaacs? How great do the similarities and specific differences between two languages, two thought traditions and their concepts have to be for them to be easily translatable? What is distorted and reframed in the process of transfer?

Clearly, this process practically invites the development of grotesque misunderstandings and the transmission of half-understood versions of ideas over generations. *Christos Pantocrator*, the shining Roman hero in the golden firmament, was one such cultural misunderstanding, an image that was only replaced by the suffering saviour on the cross once and for all after the Black Death in the fourteenth century.

In a world in which very few people could read or access manuscripts, and the vast majority only knew the Bible by hearsay, the iconography of the new religion was decisive. Here the Gospels had a clear advantage: the people of Europe could identify with the agricultural world of the Bible. It featured fields and vineyards, but also landownership and tax collectors: a divinely ordained hierarchy.

In a Europe where functional power over a large territory was often a fiction and, de facto, only extended to the edge of the next forest, because the untamed, dangerous wilderness lay behind it, monks in particular took on the task of conquering and subduing the land. Orders such as the Cistercians and the Benedictines made it their business to expand a whole network of subsidiary monasteries and agricultural estates through the wooded feudal properties, clearing the 'useless' land, ridding it of stones and breaking it open with the plough. Through shrewd purchases, centuries of donations and inheritances, the church gradually became the biggest landowner in Europe.

In spite of this campaign, civilization only penetrated the landscapes of Europe bit by bit. The biblical mandate to subdue the Earth inspired ambitious abbots and theologians, but was usually interpreted in the Augustine tradition as a command to exercise self-restraint, especially chastity, and control women. As a genuine source of power on the planet, it remained the pipe dream of theological hotheads. Its time had not yet come.

But here, in the monastic libraries, a different and important battle was being waged. The message of Christianity was radical – so radical that it was completely irreconcilable with the conception of humans in late antiquity and the attendant social practices. The Roman Empire was built on violence, and suddenly found itself confronted with a state religion whose founder had claimed to have come 'to bring a sword', yet whose message was clearly on the side of non-violence, compassion, humility and forgiveness.

There was some explaining and interpreting to do.

The Map of Misreadings

How could a compassionate, loving and forgiving saviour – who looks after the least of us, sacrifices his own son out of love and demands that one turn the other cheek – bless a society characterized by slavery, tyranny and cruel punishments, constant wars and public spectacles in which people were torn apart by animals for an audience's entertainment?

Augustine had answered that humans are simply sinful, and therefore require God's forgiveness; Rome was not built in a day. But the cruelty of these societies was not a result of individual misdeeds, it was systematic. The words of a Galilean charismatic were a thorn in the side of societies that were every bit as demonstratively violent as their heathen neighbours. Public court hearings – usually combined with a solemn Christian Mass – were an opportunity to demonstrate the legitimacy of a God-given power and render it unforgettable through spectacularly staged punishments. The ruler's will was implemented with the sword, and the face of power was marked by extreme cruelty. But if this power came from the hand of God and existed by the grace of God – why did the God of compassion give it to them?

This may sound to modern ears like an abstract intellectual game, but it had direct political implications: whereas the nobility could base its claim to power on the past, the church – which presented itself as a European power, tortured, blessed armies and sent them on campaigns – could only justify its empire with its mission on Earth.

The obvious contradiction between the biblical demand and political practice would play a decisive part in determining the following centuries of systematic subjugation: one could not simply eliminate the problematic passages from the Holy Scriptures, so they had to be interpreted intensively. Biblical exegesis became, as the literary critic Harold Bloom put it, a 'map of misreadings' – a map whose paths followed creative, systematically cultivated misunderstandings.

69

The challenge became even greater when the eschatologically tinged sayings of Yeshua ben Yosef of Nazareth, who had lived in the midst of an apocalyptic time characterized by political rebellions and doomsday preachers, were confronted with the reality of domination. Jesus preached that one should not think of tomorrow any more than do the birds in the fields, but only care for one's soul. This attitude was typical of his time, when many Jews genuinely believed that the revolt against the Romans was a harbinger of the Day of Judgement. There is little point in filling storehouses and accumulating property if the end of time has arrived and the resurrection of the dead is imminent.

One of the first great challenges for early Christianity was the fact that this apocalypse had not actually come after the death of Jesus; the world seemed to continue just as before. Gradually, the exegetes had to get used to the fact that the apocalyptic faith of an itinerant preacher would develop an everlasting institution that had not only Good News at its disposal but also palaces, fortresses and armies. The logic of power is a far cry from the burning eschatology of a charismatic. Only with the Apostle Paul did this message gain a universal meaning; now it had to be reconciled with the power of a state.

Christ, the glorious ruler in Byzantine splendour, was (viewed through the prism of Foucault) a mask of power that could follow on directly from imperial iconographic and intellectual traditions, hence the bureaucratic-looking angels in the frescoes of Ravenna or Constantinople. This Christ could preside over an empire and inspire his Caesars. Nonetheless, violence in the name of the compassionate God became a moral problem – and that was something new.

In ancient Greece (and many other societies), deadly violence was understood as a normal aspect of power. Excessive cruelty was condemned as morally wrong by Greek, Chinese and Indian scholars, but, otherwise, violence was simply part of the world's course and the necessary rituals of rule. Who exercised it and who had to suffer it was decided by the gods and the Norns, whose bony hands guided the threads of destiny.

The normality of cruelty also determined basic moral assumptions. In book 9 of the *Odyssey*, Odysseus describes his misfortune and laments that the gods are persecuting him unfairly, since he is a just man. To underline this, he tells of his journey:

'Enough; I must speak now of the fearful journeying that Zeus enforced on me when I left Troy and made for home.

'The wind behind me brought me from Ilium to Ismarus, the town of the Cicones. I sacked the town and I killed the men. As for the women and all the chattels that we took, we divided them amongst us, so that none of my men, if I could help it, should depart without his fair share.'[39]

Odysseus wants to convince his interlocutor that he is a decent and just man who shares honestly with his men and does not deserve the wrath of the gods. To illustrate his virtue, he chooses an opportunistic raid on a town that had the misfortune to be on his way, in which he and his gang laid waste to the town, murdered the men and divided up – that is, raped and enslaved – the women.

It is important to understand that, for Odysseus and his world, this was not a contradiction. Rape and murder were not immoral acts per se if they took place in the right context. *Vae victis!* (Woe to the conquered!) – thus spoke the Celt Brennus, who, the Roman historian Livy reports, had just ransacked and occupied Rome. Destiny, the gods, could bring victory and defeat even to the greatest, depending on which god was stronger, or more favourably disposed thanks to offerings, or in a better mood. To the victor belonged the spoils, but he also knew what awaited him in defeat. No one could escape the decision of fate.

Christian commentators could no longer invoke fate, since it was in God's hands: nothing happened on Earth against his will. So how was suffering possible, and how could the daily violence of the Christians be justified?

We will not wade deeply into the theological debates that developed the religious arguments to justify and reinterpret violence and cruelty in the millennium between Saint Augustine of Hippo and Thomas Aquinas in the thirteenth century. With the idea of original sin, which was passed on to each new generation since Adam and Eve's fall from grace, Augustine had laid the foundation to justify violence in the name of saving the souls of the victims. Only physical suffering, self-denial, chastisement, pain and torment could preserve the immortal soul from the flames of hell; and, compared to the hellfire, all earthly suffering was sheer bliss.

This terrible view of human life had resolute opponents within the ancient church. An Italian bishop, Julian of Eclanum, argued against original sin in a letter to Augustine:

> Babies, you say, carry the burden of another's sin, not any of their own … Explain to me, then, who this person is who sends the innocent to punishment. You answer, God … God, you say, the very one who commends his love to us, who has loved us and not spared his son but handed him over to us, he judges us in this way: he persecuted new born children; he hands over babies to eternal flames because of their bad wills, when he knows that they have not so much formed a will, good or bad.[40]

Such an act is so far removed from all piety, civilization and reason that it would not even occur to barbaric tribes, the bishop concluded. In clerical debates that increasingly strove for doctrinal unity, his voice and those of other dissenting theologians barely got a hearing, and were often actively suppressed. Any open discussion of seemingly nonsensical or contradictory dogmas was declared heresy. 'Let us Christians prefer the simplicity of our faith, which is the stronger, to the demonstrations of human reason',[41] demanded Basil of Caesarea, a Doctor of the Church, while the Carthage-born theologian Tertullian summarized the dilemma of the self-renunciation of reason in an elegant Latin saying: *credo quia absurdum est* – 'I believe because it is absurd.'

Violence and cruelty could be justified because they were fighting against sin and could save the immortal soul of the victims; for, especially in the early church, which still had to compete against other religions, the soul was always in jeopardy and the flames of hellfire licked at any human feeling. Thus, violence was also turned inwards, to the economy of individual feelings. The mere existence of the individual made it a moral battlefield; each life meant existential risk and monstrous guilt.

Origen tried to escape the surges of lust inside him through castration, legend has it. Others sought redemption in asceticism and withdrew to the deserts of Egypt and Syria to live as hermits and chasten their flesh – a process that had its own carefully cultivated and documented drama. Saint Jerome, one of the most famous of these, confessed that he had suffered from lustful and unshakable visions in the desert that were impossible to dispel: 'I often found myself surrounded by bands of dancing girls. My face was pale with fasting; but though my limbs were cold as ice, my mind was burning with desire, and the fires of lust kept bubbling up before me.'[42]

Saints and ascetics were marvellous role models, but they were of little use for ruling an empire, and the majority of those who were baptized, willingly or unwillingly, made few changes to their life, as Saint Cassian of Tangier remarked with disappointment: 'As their fervour cooled, many combined their confession of Christ with wealth; but those who kept the fervour of the apostles, recalling that former perfection, withdrew from their cities and from the society of those who thought this laxness of living permissible for themselves and for the Church.'[43] Well into the Middle Ages, there were repeated counter-movements that sought to take up the burning immediacy of Jesus' message and live by the letter of the Gospels, but these extremists were either mocked as madmen, disposed of at the stake or in dark dungeons, or, like the intransigent charismatic Francis of Assisi, quickly canonized, revered as a saint and thus rendered harmless.

Christian doctrine necessitated an enormous amount of explanation, since internal contradictions and the very worldly power of the church and the Christian rulers confronted the interpreters with ever new challenges. An entire caste of exegetes formed around the monasteries, and increasingly at the courts of nobles and in trading cities, to explain to the world why the things being done in the name of the church were God's will.

The German Jewish poet and writer Heinrich Heine, one of the most astute observers of European thought and its fault lines, believed in the mid nineteenth century that the respective legacies of classical antiquity and the Bible were in fundamental cultural contradiction – an old feud between Nazarenes and Hellenes, who are defined not by their genes, but by the structure of their thoughts and feelings:

> I could say that all people are either Jews or Hellenes, people motivated by asceticism, hostility to graven images and a deep desire for the spiritual, or people whose essential being is delight in life, pride in the development of their capacities, and realism. In this sense, there have been Hellenes among German pastors and Jews born in Athens and perhaps descended from

73

Theseus. One can rightly say that the beard does not make the Jew and the plaits do not make the Christian.[44]

For Heine, these two cultures and the war between them were responsible for the course of the European history of ideas, and it was the death of the saviour that spawned a whole new culture of subjugation: 'But only the body was mocked and crucified, the spirit was glorified, and the martyrdom of the triumphator who won world dominion for the spirit became the symbol of this victory; ever since, all humanity strives, *in imitationem Christi*, for the death of the body and for supersensual merging with absolute spirit.'[45]

History had not unfolded quite as simply as Heine feared, but he turned out to be largely right. The Nazarenes had declared the life-affirming Hellenes dissolute heretics and won the religious and ethical battle over Europe's ideas; their passion for outward and inward subjugation henceforth shaped the conception of humans and the worldview of the West. From that point on, the desiring body was a problem.

As far as the Gospels indicate, Yeshua ben Yosef had not especially hated his body. The eve of the apocalypse is not the ideal moment for cheerful, hedonistic explorations, but sexuality as such did not seem of particular moral interest to him, whereas social injustice provoked him time and again. Saint Paul, who turned the carpenter and preacher Yeshua ben Yosef into the universal saviour Jesus Christ, also brought his self-hatred into the equation, focusing it on women as temptresses and seductresses and on his own lust as the root of all evil and all sin. A less obsessive man would probably not have been able to manage his workload or reach the frenzied heights of his emotional rhetorical intensity, and the story of the sect would have ended there.

Paul guided the early parishes of the Levant and the eastern Mediterranean with virtuosity, preaching fire and brimstone. In a time of moral decadence, his burning conviction was an object of fascination. His moral obsessions suited many intellectuals with Roman upbringings, who were more interested in asceticism and stoic unworldliness because they were disgusted with the mendacity or indifference of their surroundings. His oath to gain access to salvation through the act of baptism was already familiar from Roman civil rights, which were also available to a barbarian from the provinces or a freed slave. The Roman world was ripe to be ignited by new ideas, and widely varying religions and philosophical schools fought for predominance among ordinary people and elites.

Christianity proved to be the most contagious of all religions in late antiquity, or perhaps it simply infected the right people at the right moment. With its expansion and consolidation, the idea of subjugation established itself not as the law of the strongest, but as a moral imperative conferred on humans by the Lord himself and motivated by original sin. However, this subjugation was first of all directed inwards: against one's own body, which was henceforth viewed only as a problem; against desire, which was sinful and had to be suppressed. But then it directed itself against women, who, as temptresses and daughters of Eve, and thus of original sin, had to be controlled; against heretics, heathens and Jews, who all rejected God's indivisible truth.

The subjugation of the natural environment was still beyond the technological capabilities of antiquity, but even under the Roman Caesars, there were already lasting changes to the landscape through clearing and cultivation, water management and mining. A real domination of nature, however, was so far from anything that seemed possible and imaginable that it only appeared in the thoughts of mystics and prophets.

II

LOGOS

Much greater harm has been done to the sciences by lack of ambition and by the pettiness and poverty of the projects that human industry has set itself. And, worst of all, this lack of ambition is accompanied by arrogance and contempt.

Francis Bacon, The New Organon

Landscape with the Fall of Icarus

7 Pieter Bruegel the Elder, *Landscape with the Fall of Icarus*, c.1555–68, oil on canvas, mounted on wood, 74 × 112 cm. Royal Museums of the Fine Arts, Brussels

Poor Icarus! No one cares about his fall. All that is left is a pair of naked legs, thrashing about in an undignified manner in the wake of a ship. Not even the fisherman sees him. The tragedy goes completely unnoticed. W. H. Auden writes about this moment:

> how everything turns away
> Quite leisurely from the disaster; the ploughman may
> Have heard the splash, the forsaken cry;
> But for him it was not an important failure; the sun shone
> As it had to on the white legs disappearing into the green
> Water; and the expensive delicate ship that must have seen

Something amazing, a boy falling out of the sky,
Had somewhere to get to and sailed calmly on.[1]

Spectators love high-altitude flights, but this one was short. The inevitable fall is often enough a source of *Schadenfreude*, but this one comes from too high up, beyond human concerns, which is why the attention is minimal.

It must have felt wonderful. An undreamt-of, seemingly impossible triumph over gravity, higher and higher, carried by the youthful power of his own limbs and the honey-warm air. Then the sudden panicked flapping, the powerlessness: the resistance melts away around his arms and with it the wax holding his wings together. For one horrified moment, he floats in a tumult of white, dancing feather blades, and then, with a few twitching movements like a fish on a line, he sees this blizzard before gravity seizes him again in a long tailspin at breakneck speed.

Bruegel imagines this rough landing in a somewhat comical way. The flailing, the naked bottom covered by the waves, the headlong plunge into the unplanned. The master from Brussels transposes the story from ancient Greece to the world of Flemish farmers. Without realizing it, yet with almost prophetic confidence, Bruegel captured in this picture not only the mythical past and his own present, but also the future.

But hang on; let us take a closer look at the picture first. What makes Icarus' fall even more undignified is that he is named in the title, but at best plays a secondary part in the composition. The main figure is a farmer in a red jacket and a long smock ploughing his tiny field in the foreground. He is completely focused on his work; he did not see the fall either. He is walking in ever-narrower concentric circles, and it would take a miracle of perspective or imagination for the horse to turn around at the end of the row.

But the joyless ploughman is not alone in his single-mindedness. On a ledge directly behind him, similarly foreshortened in stage-like fashion, a shepherd stands in a field and looks dreamily into the distance, albeit in the wrong direction. Perhaps his gaze is following Icarus' father, the brilliant engineer Daedalus, who had built the pairs of wings and was wise enough not to fly too close to the sun. On the right below the shepherd sits a third figure, likewise seemingly immersed in his activity: a fisherman who just seems to be in the process of casting his line. The tiny figures on the ship are indifferently facing away from the tragedy of the drowning Icarus.

Here Bruegel shows that he is an attentive reader of the Latin poet Ovid, who tells this famous story in his *Metamorphoses*. As an architect, Daedalus built King Minos of Crete the labyrinth in which the royal bastard Minotaurus is held captive: a beast half-bull and half-man. The king does not let Daedalus go either, and so the gifted master builder turns his gaze to the sky, the last place of freedom: 'The king may block my way by land or across the ocean, but the sky, surely, is open, and that is how we shall go. Minos may possess all the rest, but he does not possess the air.'[2]

Ovid's Icarus is just a boy, disturbing his father with childish matters as he constructs, and Daedalus also warns him not to fly too close to the sun. Then they flee and are observed by three witnesses, whom Bruegel also depicts: 'Some fisher, perhaps, plying his quivering rod, some shepherd leaning on his staff, or a peasant bent over his plough handle caught sight of them as they flew past and stood stock still in astonishment, believing that these creatures who could fly through the air must be gods.'[3]

But they are not immortal gods. The son loses his life, due to either childish exuberance or youthful daredevilry. The father hears his death cry, but all he can do is curse his hubris and bury his son.

However, the Flemish painter does not simply tell the story of overconfidence and sorrow; what he has to say is of concern to his own time, from which the ship and the style of the clothing come. On closer inspection, one sees motifs that add new aspects to the painting. First of all, there is the body in the forest, at the edge of the field, which is only just visible. The farmer will cut off its head with the plough-share on his next round. Dust to dust. There is a Flemish saying from Bruegel's time: 'No plough stops for the dying man.' Welcome to early capitalism.

Another detail: the money bag with the sword lying at the front edge of the field – and another saying: 'A sword and a money bag need good hands.' But the sword has impaled the money bag, as if in a sinful coitus between all the money bags and mercenaries in the world. The farmer keeps turning in his narrow field, a subjugated man who is proud of his colourful clothes. He certainly has more than the shepherd who is roaming about with his sheep. Here Bruegel reaches into the distant past – not into Latin poetry, but the Bible.

Cain and Abel are standing here. At some point the heavy-shouldered farmer will drop his plough, tie the horses to a tree stump and slay the shepherd, whose unsettled way of life is a constant and intolerable provocation for him. He will be cast out by God for it, but will never understand why. There is only one decent way to live, the farmer knows: with a house and field and plough and family and taxes and famines and military service and the blessing of the weapons in the village church. The shepherd, who roams the land like a vagabond and is not at home anywhere – this man is no longer his brother. He has long become a barbarian, an enemy and a threat to morality. But Cain's guilt remains with him, on top of the original sin of his biblical parents. The pious worker lives under the curse of his eternal guilt.

The partridge perched on a branch in the foreground, directly below the legs of the unfortunate Icarus, also has its own story, albeit a misleading one. In Ovid's version, Daedalus was punished by the gods with the death of his son because he had pushed his nephew Talos, son of his sister Perdix, off the cliffs of the Acropolis and to his death out of envy. The gods took pity on the boy and turned him into a bird, but it always flies close to the ground out of a fear of heights. The whole landscape, the whole scene with its different figures – a homily in oils? Maybe not quite. Bruegel's partridge is turned away from the spectacle as if it had nothing to do with it. This is not what triumph looks like.

The sly old painter, who was famous for hiding messages in his paintings, was not interested in edifying platitudes. For him, in any case, all these events are not a drama so much as a farce. The wax on Icarus' wings did not melt because the gods were angry, but because the boy had come too close to the burning hot sun. He was young and overconfident; the gods had nothing to do with it. They could have struck him with lightning, suddenly blinded him or made eagles attack him, but he did not even need their assistance or vengeance for his death. He had simply been foolish and his father had overestimated him – which he had to do, otherwise their escape would have been impossible. The morality tale turns into a stupid accident, with the wrath of the gods replaced by human failure. The boy Icarus was a victim of the laws of nature, both psychologically and physically.

And the ship? Bruegel's painting was created in Brussels between 1555 and 1568, an epicentre of a world that was rapidly expanding. The painter himself was well travelled for his time and had spent a number of years in Italy. In his home country, too, he was repeatedly reminded of the fact that the world was not as simple as people thought. Maritime traffic to America and Asia had begun two generations before him, and meanwhile dozens of galleys were arriving from the so-called New World, their bulging hulls filled with valuables and curiosities.

The horizon of Europeans had radically expanded, and their domain would soon grow. The port of Antwerp, along with the merchants and book printers for whom Flanders was famous, stood for this new world, and Bruegel could watch this astonishing development from a box seat. His ship (which Ovid only mentions in passing) is on the way to other horizons, undiscovered worlds and riches. No wonder it has no time to worry about a birdman in distress; there are far rarer and more profitable creatures awaiting it at the journey's end.

In a few moments, the ship will sail past the island on the left side of the picture. The island seems to consist of a single rock with a cave entrance. Is that the Minotaur's labyrinth, a Crete reduced to its essentials, even if Icarus has plunged into the sea in an entirely different place?

The scenario of this journey is a world landscape, a very old Flemish painting tradition. A landscape is not what one sees outside, for example on a mountain or a deserted plain (why would one go there anyway?); a landscape is an allegory of the world out there, with all it represents: a river or sea, with mountains, a radiant city, a wide horizon – and the sun. Is it rising or setting? Would a ship begin its voyage before sunrise? Or is it not returning from the open sea, heading for the city port? And why is the sun standing directly above the horizon when it has just destroyed Icarus' wings with its blazing power, presumably after reaching a high point?

Nothing in this painting is without ambiguity. Is it commenting on the destiny of all dreamers and visionaries, who are doomed to perish without dignity, unheeded by an indifferent world? Is it a comment on Adam's fall and original sin? Is it, as others have claimed, an alchemical programme, or rather a secret hymn to freedom in a country occupied by the Spanish and ravaged by wars? Pieter Bruegel, painter of the Tower of Babel, had a thing or two to say about the pride that comes before the fall. But is the painter referring to the arrogance of a youngster or the wise moderation of his father, whose flight from the picture's panorama has already quietly succeeded?

The painting is controversial among art historians. Was it painted by the master himself, or is it only a copy from his workshop whose original has been lost? If so, is it a faithful reproduction of the lost work? A second version, probably a second copy, shows the same scene with the farmer in the foreground and the ship sailing towards distant horizons; but this time Daedalus is also visible, a winged old man with a beard, floating heroically naked in the sky like God in the depictions that Bruegel studied during his time in Italy – but also a variation on Father Time, who was by then already familiar in Flanders from weather vanes. The work seems clumsier than the one in which Daedalus has already escaped. Probably both of them are copies of a lost original. But what copyist made it his business to correct the master's ideas?

Even the monumental, yet seemingly so modest, farmer was used as the key to a radical new interpretation. Why is he dressed so luxuriously? Why is he stepping into the freshly ploughed furrow? – something no farmer would ever do, as Bruegel, an almost fanatical collector and user of such details, knew. And why is there only one horse in front of the plough, not the usual team? Is the farmer ploughing a small piece of land perhaps an allusion to the ruler of the Habsburg Empire, Philip II, who ruled with an increasingly hard hand? Do the dagger and money bag belong to him, since he keeps the harvest and collects taxes? Then it would be something different, with the subtle message that pride sometimes – literally – comes before the fall. However much the farmer might puff himself up in the foreground, with his yellow trousers and red shirt (Philip's colours, admittedly), his mythological counterpart inevitably ends up in the water.

But no, says a different art historian, this is an allegory of transience. The half-buried body in the forest, the dagger and money bag, the farmer, the shepherd, the fisherman – these are all attributes and figures of the *Danse macabre*, the Dance of Death, in which death in the guise of a skeleton drags the representatives of different classes down with it into

84

the cold grave. If this is a moral picture, then, the moral is not a praise of moderation, criticism of bad government or a caricature of incipient capitalism, but rather a meditation on mortality itself, the vanity of all striving and work in the face of the end, which brings even the highest flyer crashing down to earth.

This work has so many messages that it has none. It is impossible to ascribe a single meaning to it because it has so many meanings, some of them contradictory. It is shimmering with possibilities, some of them very present. This is a picture about the domination of nature, going from Daedalus, who wants to change and reshape nature, via Icarus, who no longer knows his place in the great order of things, to the farmer, who is the foundation on which all power is established, yet himself dominates the animal trotting in front of him: an infinite regress of slaves and masters. The ship sailing towards the sun with billowed sails tells of domination, of a 'New World' with unimagined possibilities, but also of the fact that one's own world is growing smaller and escape is a necessity. Anyone who wants to avoid walking in ever-decreasing circles like the farmer should sign up with a ship for a voyage into the unknown, should build wings to escape from life in captivity.

One last tribute to Bruegel. 'All things begin and end in eternity', David Bowie says in Nicolas Roeg's film *The Man Who Fell to Earth* (1976), which was indirectly inspired by *The Fall of Icarus*. He plays an extra-terrestrial who has come to Earth and is corrupted by the society into which he falls. His civilization has long been receiving television and radio broadcasts from Earth and has formed an idea of the societies there. He knows that they are deeply destructive and are ruining their planet.

Bowie's figure is here on a secret mission. His own planet has almost been destroyed by wars; few have survived. His goal is to annihilate humanity and save the planet for extra-terrestrial refugees. At the same time, he becomes more acquainted with the life around him. The outsider sees the people's contradictions, their loneliness, their overwhelming emotions, and finally collapses under the pressure of his assignment and the interrogations by the secret services following him. The over-intelligent extra-terrestrial turns into a blind alcoholic who knows that he cannot save the world for anyone.

The psychedelically flickering, red-haired alien can no more evade the gravity of circumstances than the farmer or the seamen waiting in their places. There is no salvation either for the many who spend their lives under the yoke or the foolhardy who would soar up to the gods, yet find nothing there except the cruelly burning sun.

In the end it is Daedalus who escapes, beyond the picture's edge, into an unimaginable future as the first cyborg. He witnessed his son's fall and the inescapability of fate. He escapes as a broken man.

Bruegel's world marks a pivotal point in history. Like the painting, which virtuosically blends medieval imagery and Flemish painting with personal impressions, new horizons, hidden criticism, stoic observations, literary allusions, sayings and laconic irony to form a new idiom, the continent on which he lived was at a dramatic turning point that transformed the delusion of domination into a global force.

Here a question arises that, like a persistently buzzing fly, cannot simply be disposed of: the history of this delusional idea began in Western Asia and, along with the Bible, spilt over to Europe. From here, from ports like Antwerp and cities like Brussels, it would begin its global triumph in the bags of missionaries, conquerors, merchants, teachers, rebels and murderers. But why there in particular?

From a global perspective, it was a completely unpredictable development. In 1500, an observer undertaking a voyage around the world in order to identify the best candidates for world domination would not have had the slightest reason to stop off in Europe. Power and wealth, flourishing markets and great armies, cultural sophistication and cosmopolitan cities – all these things were to be found on other continents.

The Holy Roman Empire (Germany, the Netherlands, Austria and parts of northern Italy) had roughly 23 million inhabitants, slightly more than the Delhi Sultanate or the Mali Empire, but only a fifth of the most populous country, which also constituted the biggest global market and power bloc: China, with its population of 103 million, closely followed by India, the second-largest economic space, which was governed by several powerful rulers.

The middle powers of the time – France, Spain, the Songhai Empire in West Africa, the Inca Empire, the Ottoman Empire, the Bengal Sultanate and the Vijayanagara Empire in South India – had a combined population of less than 20 million. Spain, the Grand Duchy of Moscow, the Aztec Empire and England had barely more than 5 million. There is nothing to suggest that the roughly 65 million politically and religiously divided, war-torn and impoverished Europeans would set about taking over the world.

So the question remains: why Europe?

Why Europe?

It is now about 80 years since there arrived in this city of Calicut [Kozhikode] certain vessels of white Christians, who wore their hair long like Germans, and had no beards except around the mouth, such as are worn at Constantinople by cavaliers and courtiers. They landed, wearing a cuirass, helmet, and visor, and carrying a certain weapon [sword] attached to a spear. Their vessels are armed with bombards, shorter than those in use with us. Once every two years they return with 20 or 25 vessels. They are unable to tell what people they are, nor what merchandise they bring to this city, save that it includes very fine linen cloth and brassware. They load spices. Their vessels have four masts like those of Spain. If they were Germans it seems to me that we should have had some notice about them; possibly they may be Russians if they have a port there. On the arrival of the captain we may learn who these people are, for the Italian-speaking pilot, who was given him by the Moorish king, and whom he took away contrary to his inclinations, is with him, and may be able to tell.[4]

The Italian merchant Girolamo Sernigi, who sent this report to his family back home in 1499, was perplexed. Who were these strangers who had already reached India decades before Captain Vasco da Gama, in whose ships Sernigi had invested so much money? Did the Germans, with their Hanseatic cogs, even have ocean-going ships? Had anyone ever heard of a Russian fleet? But who else could these white Christians be who had become his rivals?

Sernigi, writing from Portugal, where he directed the trading post of his Florentine family business, not only received his information second-hand and through a process of multiple translations, but also two generations after the strangers, who clearly had a substantial fleet and organization at their disposal, had appeared for the last time. Practically overnight, there were no more reports about them. Other merchants had taken over their offices and warehouses, and seamen from other

88

countries repaired their sails along the docks while the carriers unloaded their cargo.

The Italian's concern was unfounded, for the mysterious naval power had long since vanished again. What he had picked up through his network of informants was a last memory of a world empire that had only existed for a few decades before being simply abandoned, suddenly and without any clear reason.

Like many recipients of information exchanged between different cultures and languages, Sernigi too was part of a chain of subtle misunderstandings and disinformation, a sort of postal 'Chinese whispers'. But the date corresponds quite closely to the historical facts, for the first fleet that actually landed on the Indian coast set sail in 1405. Even if it is true that only two dozen ships actually reached Kozhikode, the whole fleet was immensely larger. The entire horizon was filled with masts, sails and pennants, and the sea foamed from the bow waves of over 300 hulls.

Some of these vessels – perhaps 50 or 60 – were bigger than anything the world had seen before or has seen since. Based on the builders' measurements, the largest were over 120 metres long and 50 metres wide, bigger than any other wooden ship in history, longer than a soccer field, with nine masts, four decks and dozens of cannons, built to carry a 2,500-ton cargo: a floating city, armed to the teeth.

The fleet itself was bigger than the Spanish Armada and more imposing than anything that had ever been seen on the water; 28,000 men shared the bunks below deck. Along with crew and officers, merchants and translators, craftsmen and scribes, thousands of soldiers lived here for months with their horses, entire orchestras, diplomats, scholars, doctors, astronomers. On its maiden voyage in 1405, this overwhelming army reached Vietnam and Thailand, visited Java and the Strait of Malacca, then finally dropped anchor in Cochin and Kozhikode. On subsequent journeys, the fleet not only sailed to the Persian Gulf and the Red Sea, where it sent representatives to Mecca, but also to Mogadishu on the East African coast. Each time, a delegation landed, accompanied by impressive ceremonies; each time, gifts were given, tolls were exacted and deals were made, and, in many places, outposts were set up to represent the interests of this great power.

If Sernigi had heard about this unassailable presence in the Indian Ocean earlier, he would probably have invested his money elsewhere.

However, those pale rivals who wore their hair long like the Germans and had small beards like the gentlemen in Constantinople were neither Christians nor Ottomans, and even the unfortunate Italian-speaking translator, whom Vasco da Gama had evidently summarily abducted (which was no rare occurrence at the time), could not have contributed much to solving the mystery.

The commander of the enormous fleet was a Muslim from the mainland who, as a child, had been taken captive by enemy troops, castrated and enslaved. His birth name was Ma He, but he became known by the name with which his emperor honoured him: Zheng He, admiral of the fleet of the Yongle emperor (1360–1424), third emperor of the still young Ming Dynasty.

The rise of Zheng He (1371–1433 or 1435), from slave to court eunuch and finally one of the most powerful men in the Chinese Empire, was astounding. He had grown up in the southern province of Yunnan in a wealthy Muslim family; they worked in the imperial administration, and his paternal ancestors originally came from Bukhara. Both his father and his grandfather had undertaken the hajj, the traditional pilgrimage to Mecca, and one can assume that he grew up with tales of foreign lands. At the same time, the family was politically vulnerable: one of Zheng He's ancestors, Sayyid Ajall Shams al-Din Omar al-Bukhari, had been governor of Yunnan under the Mongol-led Yuan Dynasty, but the new Ming Dynasty was Han and distrusted all strangers. When the imperial army entered Yunnan in 1388, ending Mongol rule there once and for all, prominent Muslim families were persecuted. Zheng He's father died in the battle against the Chinese invaders, and the child was captured, castrated and placed in the service of the imperial court.

The young man quickly rose among the palace eunuchs and finally became a close confidant of the new Yongle emperor, who had come to power through a coup and consequently acted with either great caution or great brutality. The emperor needed capable and resolute allies, for he had great plans. They had overthrown the Yuan Dynasty only a few decades ago, but the winds of fate were uncertain; only military power could provide security.

The only real threat to his political stability had always come from the west, from the steppes, whose armies of mounted archers intermittently controlled a territory that extended from Crimea to China. The Great Wall had been erected to make an invasion by these 'barbarian hordes' impossible, or at least more difficult.

Yongle was convinced that only a massive demonstration of Chinese power would prevent the country's enemies in the northwest and on the coasts from threatening his empire militarily. He himself commanded campaigns against the Mongols in the north, and sent a Vietnamese general south with an army. But the emperor wanted to show that China was the greatest naval power too, and accordingly ordered the construction of ships whose enormous dimensions would already be enough to strike fear and terror into the heart of any foe. To command this fleet, he chose the Muslim-born Zheng He, who had a connection to the cities and people in the west.

The seven Ming expeditions were not journeys of discovery; such an immense fleet with such heavy ships would not have been suitable for that. The routes taken by the imperial ships had been used by Arab and Southeast Asian merchants for centuries, though these had travelled individual legs rather than the entire distance from China to Africa. These journeys were projections of power that were meant to bring the light of civilization to the barbarians, on the condition that they accepted the supremacy of China and paid annual tributes.

It was clearly not in the strategic interests of the fleet to capture larger territories and build up a colonial empire. The ships carried a fully equipped invasion army with them, including cavalry, and it would have been a simple matter to take over in less well-defended areas, but China lacked neither reach nor land. Zheng He's concern was to control the sea routes, and hence not only trading in the region, but also the power of regional rulers. To assert those interests, he set up a network of armed forts at strategic points along the route with barracks for soldiers and warehouses for tributes, merchandise and provisions.

In Zheng He's own words, the main purpose of the voyage was 'to collect tributes from the barbarians beyond the sea'. Cooperative local rulers thus became part of an arrangement that they had not chosen, but which also offered them protection. Those who were less willing tasted the emperor's wrath.

According to the sparse sources, King Vira Alakesvara, who ruled the small kingdom of Gampola in Ceylon (present-day Sri Lanka), tried to enrich himself on both sides by not only doing business with the Chinese, but also attacking their trading ships, or at least profiting from piracy. When Zheng He landed with an army of 2,000, Vira lured them inland to isolate them from the treasure fleet and plunder the ships. But the Chinese troops overwhelmed the resistance, destroyed the capital, killed the king's brothers and deported him to China, where he had no choice but to hope for the emperor's mercy. And, indeed, the Yongle emperor pardoned the rebellious ruler and sent him back to his country – but not without installing a puppet king and building a garrison in Kotte, the new kingdom's eponymous capital.

In return for cooperation, the new colonial masters were certainly willing to accommodate local traditions and elites. On one of his journeys to Ceylon, the Muslim admiral Zheng took a stone stele chiselled in China with him; it was inscribed with a dedication to Buddha in Persian, Tamil and Chinese, asking Buddha for protection and mercy for his fleet. Zheng's attitude to religion seems to have been pragmatic, for a different stele in southern China is adorned with his thanks to Mazu, the local goddess of sailors, for saving him from a storm. Far from adhering to a rigid politics, Zheng He showed that he was an astute politician and promising administrator of an emerging colonial empire extending to the African coast.

With its gigantic treasure fleet, the China of the Ming Dynasty dominated the Asian sea routes during the first half of the fifteenth century. There was no other naval power in Asia or elsewhere that could have dared to challenge the Middle Kingdom at sea, and within a mere two decades, Zheng He and his subordinates had built up an efficient network of vassal states and trading posts. Now it was in the Yongle emperor's hands to establish the greatest world empire on land and sea. The foundation had already been laid, and his troops, like those of the Europeans later on, had barely encountered any resistance.

But things would turn out differently: after the seventh voyage in 1433, the great fleet would never set sail again. The entire project of treasure expeditions was politically discredited, and by the end of the century it was a capital offence to travel on a ship with more than two masts. Finally, in 1525, the imperial government ordered the destruction of all ocean-going ships and confiscated all documents about Zheng He's legendary journeys. The Middle Kingdom had turned inwards. Its trading posts were disbanded and its presence in distant ports such as Kozhikode, aside from private merchant vessels, shrank to a local legend – a half-remembered rumour about strange, pale men with great ships that brought fine linen (silk) and precious brass (probably bronze or porcelain). All the Indians remembered were the long hair, the beardless moustaches and the many cannons on the ships.

What had happened?

To this day, it is difficult to say exactly why China ended its treasure trips so abruptly, especially because different factors are involved. Despite a rich stream of commodities and tributes (even including three long-necked animals known as *tsu-la-fa* – giraffes – that had come from Africa via India and became a popular motif for Chinese artists), despite the installation of rulers in distant kingdoms and demonstrations of power on exotic shores, the real challenges to imperial power were much closer to home. The Mongol threat had regrouped, and the Yongle emperor moved his capital from Nanjing on the Yangtze River – the home port of the treasure fleet and, with some 500,000 inhabitants, probably the largest city in the world at the time – to Beijing, where he was closer to the developments on the northwestern border and at the historical power base of his family.

The emperor's greatest object of enthusiasm was now the building of his new capital. To this end, he razed the old Beijing to the ground to erect a residence worthy of an emperor. The Forbidden City in the centre of the concentric city layout was a symbol of the emperor's increasing isolation from the outside world. A million workers toiled on that building site to realize the imperial dreams, entire forests were cleared to move the construction projects ahead, and an enormous canal had to be enlarged to transport the necessary materials.

But there were other factors that also doomed the treasure fleet. Zheng He and other major commanders were eunuchs who formed a power bloc of their own within the palace, and some of them were immensely influential. This repeatedly led to intrigues and even violent conflicts with the palace administration, which was dominated by career executives. The treasure expeditions were strongly associated with the eunuch faction and were increasingly caught in the political crossfire as useless prestige projects, especially in the context of the emperor's extravagant and fantastically expensive building plans. China's merchant class also wanted the trips to end, since they meant a de facto monopoly of the state on lucrative foreign trade.

The end of the expeditions came with the death of the Yongle emperor. His grandson's rule marked a political about-face: after his grandfather's ruinously costly expansionism, he concentrated on domestic politics. He reduced the excessive state powers introduced by Yongle, lowered taxes – which had kept rising thanks to the construction projects – promoted scholars rather than generals, and ushered in a new era. The treasure voyages, once his grandfather's pet project, were a thing of the past. A single expedition took place during his reign. One last time, in 1433, Zheng He returned with the ambassadors of eleven tributary kingdoms so that they could prostrate themselves before the new emperor. These realms included Malaysia, Kozhikode, Cochin, Ceylon, Dhofar, Aden, Hormuz and Mecca. The admiral died two years later (some sources claim that he already died on the homeward journey, and only his shoes returned to Nanjing aboard his flagship) and with him the age of treasure voyages – and an immense, unwritten chapter of historical possibilities.

Why Europe? As the Ming treasure voyages show, it is historical coincidence that comes into play first of all. The colonialism of the so-called occident was only able to spread because China had voluntarily given up its control over large parts of the sea routes and trading hubs of Asia and Africa, just when the Europeans had finally caught up with technological developments in Chinese shipbuilding, such as compass needles, three-masters, outside rudders and ocean-going hulls. Within a few decades, Portuguese caravels (mere nutshells compared to the gigantic treasure ships) filled the vacuum in the Indian Ocean, taking over parts of the Chinese infrastructure and employing Chinese maps, which showed the routes all the way to Mogadishu with exact bearings, travelling times and astronomic aids to orientation.

But this vacuum alone does not explain how the small, initially second-tier continent of Europe could, within three centuries, become strong enough to carry its theologically founded delusion of subjugation into the whole world – considering that Chinese society was vastly superior to the barbarians from the distant north in almost every way.

The Chinese peasants had fuller stomachs and lived longer than their European cousins, thanks to immense irrigation networks that were set up over centuries and made it possible to produce rice for a population that was huge by the standards of the time. Chinese inventiveness had brought forth gunpowder and the cannon, printing with movable type, paper, banks and credit systems and the compass, long before they were known or used in Europe. A Chinese emperor could find a million men for his palace, build an invincible fleet, divert rivers and command the largest army in the world. No other ruler on the globe held such power.

In addition to this highly developed practical culture, China could look back on a long philosophical tradition, starting roughly at the same time as that of ancient Greece, namely in the fifth century BCE, in a time when both places were characterized by wars and insecurity.

Maybe it was partly in response to this instability that Confucianism was intensely concerned with social harmony, the power of traditions and rituals. This school of thought was in constant tension with that of the Tao, the 'way'. The Tao is the flow of the universe and all living things, an ever-changing nature beyond all conceptualization that only reveals its continuity in flowing. The wise person asks what the nature of the Tao is and seeks to live by it; the foolish one opposes it and lives in ignorance, greed and envy, violating the law of all that lives.

Here we should note that the complexity, the internal debates, the spaces for interpretation and the iridescent possibilities of the Chinese intellectual tradition are among the great treasures of human history, but that I am forced to rely on second-hand sources because I do not have direct access to the play of meaning in Chinese thought. But Taoism and its relation to Confucian thought are so central to the idea of dominating nature that I must at least offer an outline here.

The doctrine of the Tao is reflected, in all its provocative elasticity, in the life and writings of Zhuang Zhou (c.369–286 BCE), known to his contemporaries as Zhuangzi or 'Master Zhuang'. His penetrating scepticism is evident in the allegorical tales in which he clothes his teachings, always with the resigned smile of someone who cannot be surprised, yet has still not forgotten how to love. His most famous story concerns himself and his fatigue:

Once Zhuang Zhou dreamt he was a butterfly, a butterfly flitting and fluttering around, happy with himself and doing as he pleased. He didn't know he was Zhuang Zhou. Suddenly he woke up and there he was, solid and unmistakable Zhuang Zhou. But he didn't know if he was Zhuang Zhou who had dreamt he was a butterfly, or a butterfly dreaming he was Zhuang Zhou. Between Zhuang Zhou and a butterfly there must be *some* distinction! This is called the Transformation of Things.[5]

The interplay of identity and change fascinated the thinker, who had no illusions about human nature:

> With everything they [humans] meet, they become entangled. Day after day they use their minds in strife, sometimes grandiose, sometimes sly, sometimes petty. [...] Joy, anger, grief, delight, worry, regret, fickleness, inflexibility, modesty, wilfulness, candour, insolence – music from hollow pipes, mushrooms springing up in dampness, day and night replacing each other before us, and no one knows where they sprout from.[6]

Unrestrained emotions play on humans like organ pipes, and their emotions are no more significant than the wind, which blows across hollow pipes and produces notes through them. Only the philosopher who learns to distance themselves from this involuntary agitation and look behind life's facade will recognize that true happiness lies in surrendering to the rhythms of nature and the irretrievability of the moment, and relinquishing control of the outside world. Ultimately, it matters little whether Master Zhuang is a man or a butterfly, or whether he knows which of the two applies; what matters is only whether he understands his condition as a dreamer, and can thus liberate himself from suffering through ignorance, for ignorance reduces even the rich and the powerful to objects of their own greed: 'For to possess territory is to possess a very great thing, and one who possesses so great a thing cannot do so by being a mere thing himself.'[7]

For Confucius, as well as Zhuang Zhou and his Taoist students, it was important to live in accordance with objective principles, even if they described these principles in different ways. In this sense, the Western cliché that Confucian societies are concerned above all with social harmony, not individual happiness, is not entirely inaccurate, though it does pass over two millennia of discussions and historical developments such as the confrontation with Buddhism.

The intellectual traditions of 'Subdue the Earth' and living in accordance with the flow of the *qi* and the ineffable way of the Tao could, at first glance, scarcely be more different. In truth, however, they share an important aspect that lies outside of these ideologies. The two civilizations thought very differently about the world around them, but they both approached it in an amazingly similar way.

The environmental historian Daniel R. Headrick describes how societies at similar levels of technological development deal very similarly with their natural resources. First comes the extermination of large land animals (such as aurochs in Europe, mastodons in North America, lions in Mesopotamia and tigers and elephants in China); then, with the growth of cities and the beginnings of metalworking, comes the insatiable need for wood and the clearing of entire wooded areas, to say nothing of the changes to nature through animal breeding and the transformation of entire landscapes. The Chinese government was working systematically on this as early as the Song Dynasty:

> It published instruction manuals, provided seeds, and offered low-interest loans and tax rebates to farmers. It established military colonies and settlements on state lands for refugees and landless peasants. Under the direction of landowners and businessmen, migrant farmers undertook the reclamation of the Jiangnan, transforming that vast region of salt-marshes into the most densely populated and productive farmland in China.[8]

China's intensive land cultivation and exploitation of natural resources already before 1500 show that social history does not necessarily determine how civilizations treat their natural surroundings. Chinese philosophy built on an ancient, important tradition of living in harmony with the flow of nature, and yet barely a river was left in its natural bed.

However, highly organized and effective agriculture was also a weak spot for China as a great power. The rapidly growing population of the Ming Dynasty could only be fed through intensive cultivation of rice and other agricultural products, which necessitated building canals across great distances to water the fields and double the harvests. The first irrigation canals in China were made almost as early as those in Uruk. The Lianghzu culture on the southern coast erected a city with canals and palaces surrounded by irrigated fields as early as c.3000 BCE, and many of the canal systems of the Ming era originated from earlier dynasties. At the same time, the canals constantly had to be re-dug and cleaned, and floodgates, reservoirs, banks and dams had to be repaired and maintained.

In peacetime, the excellent administrative apparatus functioned and planned the necessary tasks. In northern China especially, however, this system was very prone to attacks from without or rebellions from within. A broken dam could flood entire rural areas and make them unusable, and a diverted river, or even just years of insecurity, could delay the repairs: millions would be threatened with starvation if the complex network were torn apart. During the conflicts with the Mongols, who repeatedly attacked from the West Asian steppes, such interruptions and the resulting famines occurred time and again.

The supply shortfalls of a huge population and the constant threat on the periphery, as well as through regional uprisings, illustrated a further problem that large empires often experienced (and still do): a considerable part of their military and strategic energies and financial resources are absorbed in the interior to maintain rule. A large empire has the means to build a fleet of unprecedented size and send it around the world, but the pull of events and the decision-making power of individual persons could sometimes reverse priorities completely.

These weak points may have prevented China from permanently exposing itself by expanding its realm of influence, which was already very large. Communication channels and transport routes were long – in some areas, too long for a central government to control them effectively. But none of the kingdom's weaknesses was remotely as grave as the problems resulting from its indomitable strength.

With or without its armada, China was the greatest power in Asia. Except for the Mongols, who kept trying their luck, there were no serious adversaries in the region threatening to attack China or surpass it economically. During the early Ming Dynasty, the economy boomed and the population doubled from 100 to 200 million in the course of the sixteenth century. But this also removed any incentive to further develop technologies that had been working superbly for generations, or to introduce new ones.

One example of the disastrous chains of coincidences that sometimes write history is the development of firearms, which China had already been using since the ninth century. The only serious outside enemy were the hordes of riders from the steppes. A trained rider could fire several arrows per minute from a galloping horse in an attack that might only last seconds – thus, the entire troop company could send out a hail of arrows and quickly retreat. Such moving targets were difficult to hit with a Chinese fire lance, or some other early portable cannon or musket, and the shooters were also practically defenceless while reloading. It was not in the interests of the Chinese land army, then, to develop these spectacular yet impractical weapons further, while in Europe, an entirely different kind of war was emerging in which faster, more accurate and mass-produced firearms were in use. When the two powers came face to face 200 years later, this deficit would prove fatal.

Technology and the Burden of Empire

China was not the only empire whose sheer size made it impossible, or at least difficult, to find a global stage for its ambitions or fundamentally change its own technologies and practices.

The Ottoman Empire likewise had the organization, the technology, the geographical knowledge and the expansionist urges to project itself further into the world. What it lacked was a port with access to the open sea – leaving aside Aden at the southern tip of the Arabian Peninsula, far from Constantinople and difficult to defend if it came down to it. The naval power of the Ottomans lay in the Mediterranean and was based on its fleets of galleys, which could be employed there to lethal effect, but were completely unsuitable for voyages on the open sea. In addition, the Ottoman fleet was effectively shut in by the Strait of Gibraltar. Without an appropriate entrance to the open sea, however, the Ottoman Empire could not hope to build up a further network of trading posts and garrisons in order to consolidate its global influence.

In addition to the disadvantages of the empire's geographical location, its sultans – much like their peers on the dragon throne – were constantly occupied with crushing regional uprisings and fending off invasions from the same Asian steppes, whose westernmost foothills lay in Ottoman territory. The enemies of China were also those of the Ottomans, and, strangely enough, the battle against them likewise seems to have held back the sultan's troops in their technological evolution, since firearms were not yet sufficiently effective against mounted armies.

Nonetheless, the Ottoman Empire remained an immense presence in Eurasia. Suleiman the Magnificent (c.1495–1566) was obsessed with the idea of world domination, even if Islam itself, the Qur'an, the Hadiths and the scripturalists did not emphasize the domination of nature any more than their Christian neighbours did in their own theological discussions at the time. He had an inscription placed in the white stone above the entrance to the Süleymaniye Mosque: 'Conqueror of the lands of the

Orient and the Occident with the help of Almighty God and his victorious army, possessor of the kingdoms of the World'. Under Suleiman, the High Porte's sphere of power was dramatically expanded: in 1521, he conquered Belgrade and thus Serbia, then Rhodes and Hungary; in 1529, the Ottoman army reached the gates of Vienna but had to retreat, thwarted not by Christian soldiers but by the onset of winter.

In fact, however, history itself was working against the Ottoman Empire. Just as the Ottomans were taking increasing control of the immensely profitable trade in the eastern Mediterranean, it suddenly lost significance. The Atlantic and the route around the Cape of Good Hope created an enormous new hinterland for European states and an unprecedentedly profitable trade in spices, raw materials, consumer items and slaves to other continents. It is no coincidence that all the countries which would establish great colonial empires – Portugal, Spain, Great Britain, France, Belgium and the Netherlands – controlled Atlantic ports. The Ottoman fleet could only reach the ocean through the Strait of Gibraltar. The sultan's admirals controlled the Levantine waters and the land routes to Asia, but the Atlantic trade deprived the eastern Mediterranean of its strategic position, which was also felt in Venice.

The Ottoman naval forces were powerful and feared, but were almost entirely limited to the Mediterranean. The land army relied heavily on cavalry units, since most of their enemies, from Arabia to the Central Asian steppes, were mounted nomads who attacked with a hail of arrows fired rapidly on horseback.

Yet this would prove detrimental to the power of the sultans, for, although firearms, especially huge cannons, had long been in use by the Ottoman troops, they were – as in China – almost useless against fast mounted units. The use of muskets was almost impossible for riders, since they could not reload in the saddle. So the Ottoman strategy relied on cavalry with lances, bows and arrows, long after firearms had been widely adopted in Europe and rapidly refined.

European wars took a different course. Here, power was controlled from the cities, which meant that sieges were an especially important instrument, and their success depended on artillery, which was continually perfected. But the battles between the European powers were also different from the conflicts between the Ottomans or Chinese and their nomadic assailants.

Foot soldiers were cheaper to train and more numerous than cavalry. In the geography of the European landscape, subdivided into small elements with its forests, valleys and rivers, the armies fought mostly on foot in closed formations, which could be very effectively attacked with untargeted salvos from a wall of firearms. The musketeers, who were almost defenceless while reloading, had to be protected from cavalry attacks by men with long lances, which made them ideal targets for cannon fire.

Early arquebuses or muskets were already being fired at the Battle of Agincourt in 1415, but the battle was won by archers. Some 200 years later, when the Thirty Years War broke out, muskets and cannons that were fast, accurate and robust had become decisive for the war, and no commander could afford to fall behind in the competition for technology, tactics and knowledge.

At the same time, Europe's next rivals made a number of fatal decisions that would set them back once and for all. The religious authorities had always been sceptical about the acquisition of knowledge and new technologies – not only from the West, but also from their own production. Twice, in 1485 and 1515, they enforced an official ban on printing books or other documents in Arabic or Turkish, so that the only works in these languages would be produced in Rome, Venice and northern Europe, while in the Ottoman Empire every document had to be copied by hand. The first official Arabic and Turkish printing press in Constantinople opened its doors in 1726.

The important scholars and academics did not fare much better. As early as the eleventh century, the polymath Abū Rayhān al-Bīrūnī had travelled to India and written about the societies and religions there, calculated the circumference of the Earth to within 40 kilometres, translated Euclid into Sanskrit, predicted eclipses and reflected on the gravity and density of physical substances. Things were considerably more difficult for his successors.

Taqi ad-Din Muhammad ibn Ma'ruf (1526–85) built astoundingly accurate clocks two centuries before the Englishman John Harrison. He also developed an early form of steam engine. Sultan Murad III generously sponsored his research because he expected this astronomical work to offer a way of predicting the future from the stars, just as European rulers of the same period became patrons of alchemists and kabbalists to acquire gold and insights into the alphabet of the universe. Ad-Din was also the director of the observatory he had founded in Istanbul, where he saw a comet in 1577 and interpreted it as a good omen for the sultan's military campaigns. When his troops were subsequently beaten, Murad III withdrew his support for the extravagant project, and the observatory was demolished in 1580.

The great admiral and cartographer Piri Reis, who drew a remarkably accurate map of the Mediterranean and Atlantic, including the coast of America, in 1513, worked exclusively with second-hand information from European cartographers and published maps, including a lost map of the American coast that had belonged to Columbus, or so he claimed. He himself sailed around the Arabian Peninsula, but never saw any more of the world that interested him so intensely.

When one asks why European ideas spread and established themselves all over the globe, perhaps the most important part of the answer is that chance was on the continent's side. Just when European ships became able to circumnavigate the Cape of Good Hope and progress as far as Asia, the established power had withdrawn and abandoned all imperial ambitions. Other potential rivals like the Ottoman Empire lacked any suitable ports or other strategic priorities. The peoples of the Americas had not developed any deep-sea navigation.

Geographically, too, Europe had certain advantages to offer. The coast of the continent from Scotland to Spain had free access to the Atlantic, which gave the rulers and merchants of several countries the possibility to seek kingdoms and treasures on other continents. An important part of the answer is still missing, however; while one can easily argue that it was precisely the regional strength of China and the Ottoman Empire that led to their decline, one can also observe that the weakness of Europe was the starting point for its global power.

With the collapse of the Roman Empire, Europe lost its unity. The post-imperial parts of the continent with a strong Latin influence drifted apart; Christianity had only established itself in parts of Europe, and none of the rival powers was strong enough to win out over the others. No European emperor could assert his will over the others alone, or move the political orientation of the empire in a completely new direction; the power of the archdukes, the nobility, the kingdoms and the clergy was much too great. Europe had no ruler. It remained a collection of competing minor powers on all political levels, a constant struggle of all against all – and this was the beginning of Europe's rise.

Surrounded by fickle allies and sworn enemies, war was a constant reality for every European state. No ruler could afford to be a pacifist in the face of the territorial greed of his neighbours. Every county, every free city and every princedom had to fend off invasions or summon troops and money for the conflicts of its allies. A life without war on the horizon was unthinkable.

Military strength depended firstly on the size of the population and the resulting tax revenue (war was always expensive, and mercenary armies could quickly change sides if they were not paid), and secondly on other advantages that made a difference in the never-ending arms race: whoever had the largest trading cities profited – not only finan- cially – from the exchange of commodities, news, technologies and ideas; anyone who possessed natural resources could become a valuable ally or trading partner; those who married wisely could expand their sphere of power; whoever had the better ships, the most professional artillery, the more effective banking system and the best engineers and artists working for them was ahead of their rivals – perhaps narrowly, but sometimes decisively.

Between 1400 and 1500 alone, there were several overlapping and successive wars in Europe: the Hundred Years War between England and France, the Ottoman–Hungarian wars, the Hussite wars in Central Europe, the battle for dominance in northern Italy between Milan and Venice, as well as a repeatedly erupting conflict between Teutonic knights and Prussian nobles on one side and Polish-Lithuanian rulers on the other side, a four-year revolt in Ghent, the Wars of the Roses in northern England, a civil war in Catalonia, a confrontation between the Venetian fleet and the Ottoman Empire at the Battle of Lepanto in 1479, skirmishes between North German Hanseatic towns, British battleships and piracy on almost every coast – to name only the most important ones. Not a year passed without fighting in at least one place in Europe.

Here too, war is the father of all things, and the arms race expedites technological development among European powers. Armies with a large proportion of cavalry (as was typical in China, the Ottoman Empire and the Mughal Empire) were unsuited to the more subdivided geography of European marching routes and battlefields. Furthermore, mounted soldiers were considerably more expensive to train, arm and maintain than infantry.

Thanks to their long pikes, infantry units could keep advancing enemy soldiers at a distance very effectively by creating a forest of rigid lances with flashing steel tips. To achieve this, they marched in close formation and formed a hedgehog with spines up to 6 metres long. Firearms had a distinct advantage in such situations. They were not accurate and took a long time to reload, but the compact mass of bodies made an ideal target for cannons and arquebuses, and the explosions and deaths caused from afar had an enormous effect on morale.

In battle, especially during sieges, artillery became a key technology, enabling an attacking army to destroy a city, fortress or army from the air without risking its own soldiers, while specialist sappers dug ditches and tunnels and attempted to place explosive charges under the fortress walls. Engineers, metallurgists, mathematicians and other specialists became sought-after and highly paid experts who calculated the trajectories of projectiles, planned fortresses, improved weapons, trained troops and secured logistics. These scientifically minded people were often key figures in cultural development. Leonardo da Vinci was one of these military technologists, and René Descartes was trained in the calculation of launching angles and trajectories as an artillery officer before going on to write philosophical works.

The Justification Industry

War not only advanced technological developments. Especially after the Reformation, it fuelled the theological industry of justification, which had shaped European culture since late antiquity.

During the debates and councils of late antiquity, the official theology of the church had been temporarily stabilized and successfully unified. Time and again in subsequent centuries, there were movements such as the Cathars or Albigensians that opposed the official doctrine, but most of them were wiped out by the armies of the pope and his allies or neutralized by the Inquisition. In the fifteenth century, however, the centrifugal forces of the debates and political interests finally became too great and the unity of the church collapsed.

After Martin Luther's rebellion within the church had divided believers, the one indivisible truth of God was also shattered and now had several churches, interpretations and armies. Because theological dogmas were always tied to power, however, the quarrel over religious truths soon became a *casus belli*. Each side had to prove not only that it was stronger, but also that it alone embodied God's will; professional propaganda became indispensable in all areas.

The wars of religion that bled the continent dry from the sixteenth century onwards mingled political interests with religious convictions. Unlike Odysseus, however, theologians had to justify the slaughters carried out by mercenaries and their armies, which depopulated entire swathes of land, in such a way that they did not stand before God as sinners, but rather as the army of the Lord, who asserted his will and his mercy with the sword.

God's truth had become a battle cry. During the Crusades, the wrath of the Lord was directed at the 'unbelievers' who defiled his tomb and occupied his Holy City. This had been an easy argument to make. After the Reformation, however, the unbelievers who had to be converted or exterminated were brothers who prayed to the same god and thanked the same saviour for his mercy. Augustine had already expended considerable rhetorical energy giving a theoretical foundation for violence in the name of God and his gospel. Now that Christians were murdering other Christians in the name of Christ, however, the pressure to justify one's actions massively increased.

The justification industry came about not through a coordinated plot or conspiracy, but simply through the feeling that it was necessary to resolve certain immanent or even visible contradictions that emerged especially brutally in the wars of religion – in a climate in which there was no freedom of opinion and political opportunism prevailed. How could all that be God's will? And how could the murderers and profiteers and warlords who destroyed harvests and wiped out villages year after year be good Christians?

The intellectual battle for God's blessing raged in Europe until the Enlightenment and beyond. Theologians and historians, jurists and philosophers (and later also archaeologists, anthropologists, zoologists and so on) joined the debate and, remarkably often (with exceptions such as Machiavelli, who described the whole thing as a game), came to the conclusion that, for irrefutable reasons, their employer was on the right side of history.

Of course, the necessity of justifying violence arises in all cultures, but the gulf between the brutal practices and the religious dogmas in Christian countries was wider than in contexts where the violence of the powerful and the victors was not a priori a moral problem, but subject to the law of the strongest, albeit limited by rules of proportionality and appropriateness. Christian societies, however, had to prove to themselves and others that systemic violence and organized hatred on all levels were genuinely the will of an all-merciful, all-knowing and good God.

This competition for God's blessing or the best arguments for it was no less important than the race for faster and more accurate firearms. No other society in the world, with the possible exception of China, developed such a wealth of professional explorers, explainers and interpreters, of propagandists and poets, who sang the praises of their side using all the analytical, rhetorical and aesthetic means at their disposal – for there were many alliances in different conflicts, so there were many songs of praise to sing, frescoes to paint, works to write and leaflets to print.

When one speaks of early printing, it is often forgotten that producing books was not the only work done by printers, and perhaps not even the most important. Books, whose production involved typesetting, printing, drying and binding each page by hand, were extremely expensive until the seventeenth century. But leaflets adorned with illustrations, often written in rhyme and printed on one page could be cheaply sold and

8 Unknown artist, 'Augusta Angustiata, a Deo per Deum liberata'. Stiftung Preussischer Kulturbesitz, Ident. no. 14136034. Photo: Art Library, Berlin State Museums

easily hidden or smuggled. Satires, horror stories or even suggestive pictures unleashed an incredible political impact. In the propaganda war between confessions, pamphlets served as weapons that could be employed quickly and flexibly to spread disinformation and demoralize the enemy – or to keep one's own people in line.

During the wars of religion, which ravaged entire landscapes in the name of the Lord and left behind deserted strips of land strewn with corpses, there was a constant pressure to prove by any means that the Lord was on one's side, no matter what atrocities were committed under his benediction.

King Gustav Adolf of Sweden was the most important Protestant commander in the Thirty Years War. In an anonymous leaflet from 1632 (figure 8), he is depicted as 'The man who can help', having just liberated Augsburg from a siege by Catholic troops. The warlord is on the right hand of the picture, with the city symbolized by a lamenting widow beside him and a church looted and desecrated by Catholics on the left. God's name appears above the actual city in a cloud, shining on the Swedish king: the instrument of the Lord, who must rule by fire and the sword to enforce his will on Earth, against the perverse will of the other side, which had become not only an enemy but also a traitor.

The Age of Iron

Military and economic competition, access to the open sea and the theological justification industry changed European societies and triggered an inescapable contest for political and religious dominance. These particularities were brought out especially strongly when a global climate crisis, the Little Ice Age, put societies under pressure.

The 'Little Ice Age' is a meteorological event that has not yet been fully scientifically explained, but is extremely well documented. It refers to a period of cold that reached its apex somewhere between 1560 and 1685, with an average temperature drop of 2 degrees Celsius.[9] This meant a maximum drop of 8 degrees in seemingly endless winters and short rainy summers, with spoilt harvests and impending famines. The crop cycle was shortened by three weeks, the maturing time came too late, and farmers had to watch time and again as the wheat in the fields rotted away.

An analysis of ice cores, tree rings, plant sediments and other natural indicators, as well as human documents such as harvest journals, weather records, letters and diaries, makes it clear that this crisis hit several continents with equal force, but a productive comparison is rendered impossible by the great discrepancies of emphasis between sources. In the case of Europe, there is an overwhelming abundance of documents that allow a detailed reconstruction of the crisis and its economic, political and philosophical effects. This is more difficult for other areas; there is only scant and indirect historical evidence of the effects of the crisis on the societies of America, Australia and large parts of Asia – again with the exception of China, where the political and demographic changes of the time are exceedingly well documented.

The China of the Ming Dynasty in the early seventeenth century was hit brutally by the climate catastrophe. In Jiangxi Province, orange trees that had borne fruit for centuries were cut down because it was no longer warm enough. What was much worse, however, was the situation in the northwest, which produced a substantial part of the millet and wheat harvest. The long, hard winters and short summers were compounded by attacks from Mongolian riders, who destroyed canals and dams or prevented urgently necessary repairs. The effective, but disturbance-prone, irrigation system was interrupted time and again, leading to crop failure. Starvation, uprisings, tax shortfalls and a consequent decline in investments in the infrastructure exacerbated the situation.

There was another, very individual reason for the lack of energetic policies. The Wanli emperor (1563–1620) was only ten when he ascended to the throne, so government affairs were mostly managed by his grand secretary Zhang Juzheng, a vigorous reformer of such disparate branches as agriculture and central administration. He made few friends in the course of his career, but introduced important innovations.

After Zhang Juzheng's death in 1593, the young emperor himself took over the reins and rebelled against the man who had become his political mentor by reversing many of his reforms, then escalating a conflict with the palace administration because he wanted to name his third son, not his first, as his heir to the throne; he lost the battle. This quarrel resulted in the Wanli emperor withdrawing completely from government affairs from 1600 on; he no longer made important decisions, appointed officials or set any political course. The emperor of China was on strike, a strike that continued for twenty years.

The situation was also tense in Manchuria, in the northeast of the empire, because heavy snowfalls prolonged the sowing period until spring, resulting in failed harvests and starving livestock. When the striking emperor's administration insisted on the full payment of tributes, the powerful clans of the region rebelled and drew the emperor into a war that would incur heavy losses for him. The extreme weather continued: typhoons on the coast, flooding of the Yellow River and drought in other areas. Time and again, harsh winters punished rural society and weakened the population, afflicting the country with constant plagues. According to some estimates, half of the Chinese population in the seventeenth century died from starvation, epidemics or war.

By 1644, the government of the last Ming emperor was dramatically weakened, and it was ultimately toppled by a peasant revolt. This revolt began out of desperation over the crushing tax burden imposed by the central government, which often used force of arms or imposed severe punishments to collect its taxes. One of the countless starving peasants was a man named Li Zicheng, who was put in the stocks in the marketplace because he had been unable to pay his taxes. That was the last straw: the furious crowd freed Li Zicheng and brought him to the mountains, from where he commanded first a gang of men with clubs, then an army of rebels with tens of thousands of soldiers. The imperial army was powerless against the rage of the peasant army. When the news reached the Forbidden City that Li Zicheng and his men had taken Beijing, the last emperor hanged himself on a tree in the imperial garden.

The Little Ice Age was a decisive factor in the collapse of the Ming Dynasty, even if there were also other important elements. The decision of the Wanli emperor to reverse important reforms and withdraw entirely from politics left a vacuum in the government, which became embroiled in battles between palace bureaucrats and eunuchs, and essentially ceased to function. A rigid administration lost control of the country, and was ultimately overwhelmed completely by the combined strain of the agricultural crisis, the invasions from the northwest and the civil wars that were breaking out.

It is fascinating to see how and why Europe found a different answer to a very similar initial situation over time. Here, too, the populace had been hit hard by protracted winters and failed harvests; here, too, there were uprisings; here, too, there was starvation and war, especially in ravaged Central Europe, where, similarly to China, probably half the population fell victim to these conditions. The winters were so cold that entire armies could ride over the frozen Danube and the icy Rhine during the Thirty Years War, and new neighbourhoods grew in London during the cold months. Stories were told about birds falling from the sky, frozen in mid-flight, and riders toppling off their horses as lumps of ice. We do know that the ration of wine for soldiers sometimes had to be sawn off large blocks.

The first responses of the Europeans to this obvious threat were very medieval, as one would expect. There were processions, prayers of supplication, sermons; relics were carried around churches and to glaciers; men in long robes flagellated themselves in the streets; and every bad harvest was followed by a wave of witch trials through Central Europe in which women (and occasionally men) were accused of cursing the crops and making the livestock sick – and fornicating with the devil, which was part of the package, not least to arouse the public's interest.

Not surprisingly, these measures were unsuccessful in changing the weather. Harvests continued to fail and the price of flour and firewood doubled each year. At the same time, purely by coincidence, the continent was substantially better at finding long-term solutions: in a decentralized, sometimes anarchic fashion, by failing and failing better, and through empirical observation and a certain openness to innovations that could lead to a competitive edge.

The first genuinely constructive measures in Europe consisted in a change of agricultural methods and products under the guidance of botanical experts, who put the results of their published series of experiments up for discussion. Increased long-distance trading in wheat and other foodstuffs helped to balance out shortages after crop failures and strengthened the exchange of commodities, people and ideas, as well as the need for reliable news and legal security.

The pressure to reform had risen everywhere because of the agricultural crisis, but it could be ignored for decades in a realm with a single ruler if this ruler proved unsuitable. In Europe, too, there were several inept rulers at that time, but only a few days' travel away, rival powers were developing further, in some cases poaching the brightest minds of a stagnating court – a fate that befell the Spanish Habsburgs in the sixteenth century, to name one example.

At the same time, Europe also profited from sheer good luck. European farmers suffered from the weather and the constant wars too, but their agriculture was based on cereals rather than rice as a staple, which meant that it was a system of smaller parts rather than a centrally administered irrigation system, and damage was thus more locally restricted. Even a field in which the corn had rotted in a bad year, or been burnt by a marauding army to the despair of the rural population, could yield again the following year. A field planted with the new potato even dealt with the cold better, and could not fall victim to flames.

The Little Ice Age necessitated radical changes, and amid the intense competition between European powers, the effects of new practices soon became apparent. By the first half of the seventeenth century, there were barely any more processions or Masses against the cold, but an extensive trading network had been built up, a living network of readers and a strengthened urban middle class consisting of merchants and lawyers, tax collectors and doctors, rentiers and entrepreneurs, that now began agitating to assert their own interests.

Different elements of an answer to the question of what constituted the specifically European path of history have now become clear, but we still lack a final answer that would explain everything. The geographical position of the continent, with its long ocean coasts, was just as important as the existence of large forests, natural resources such as iron and coal and the intense competition among individual countries and kingdoms.

From the start, a central element in the question of European dominance was the creative destruction of war, for which European states, in contrast to China and Japan, often spent over two-thirds of the state budget. This stimulated technological, scientific, economic, administrative and even (propagandist) artistic evolution, and creative minds were consequently sought after and well paid all over the continent.

Centrally ruled large empires such as China and the Ottoman Empire, which also came under pressure in the Little Ice Age and pursued isolationist agendas for political reasons, defended the authority of their ruling houses by emphasizing religious and cultural purity and timelessness. This also led to a general isolationist attitude, with book burnings, printing bans and destruction of machines, as well as the educational restrictions repeatedly imposed during that turbulent century.

In the European context, however, cultural purity was a cultural fiction from the start. Religious purity was already controversial, and was the object of intense discussions that set off an endless flood of books, but any discussion and any trading house needs reliable scribes and bookkeepers, as well as quick minds in all areas. Arguments about the true faith could never be restricted to faith, and always drifted towards social and political questions, discussions about justice, equality, human dignity and other dangerous ideas. Once a conversation has begun, however, it cannot simply be shut down again, even if the original speakers have long since been robbed of their quills, tongues or immortal souls by the Inquisition or an impatient sovereign.

New technologies inevitably brought new thoughts and heretical talk, which, in printed form, were more lasting and subversive than remarks between friends. The continent-wide discussion that had begun as a confrontation between theological positions and moral justifications for systemic violence developed a dynamic that separated it from theology and posed universal questions. In so doing, it questioned the power of the powerful and became the declared enemy. New theologians, philosophers, historians and artists were trained to answer these questions in the only right way, and they too, along with their students, were not immune to the absurdity of their tasks and to their own critical instinct – so the game started all over again. The legitimacy of the church died partly because of the zeal of its most ardent defenders.

Much of this intellectual creativity and the uncounted riches were invested in the absurd aristocratic obsession with military poses and glorious victories, but the insights that became the foundations of modern science, as well as many of the artworks that are still admired today, would have been unthinkable without this obsession with deadly technologies for war. This was no use to the peasants who were raped, robbed and left to starve by marauding troops, but it accelerated technological change.

The other side of the state of war and constant violence was the need for a Christian society to understand itself and its interest as virtuous and pleasing to God. The justifiers, whitewashers, founders and cultural window-dressers of European societies played an essential part in societies where violence had become a fundamental moral question as well as a part of daily experience. Thus, violence and oppression in the service of moral justification were elevated to the only correct and God-pleasing way of acting and thinking: a breathtaking and unparalleled institutionalization of hypocrisy.

Monsieur Grat and His Master

Did he love him? If we are to believe contemporaries, Monsieur Descartes and Monsieur Grat were inseparable and spent long hours together – and entire nights in the philosopher's overheated bedroom.

As the name indicates, 'Mr Scratch' was not a human companion. He accompanied his friend on four legs, for he was a dog whose master had a special relationship with animals. Descartes was intensely interested in animals, albeit not always in the friendliest way. The name of his dog may already be a sign that he was especially curious about reflexes and involuntary actions, and explored them himself. In one of his works he describes the vivisection of a dog's heart, which he feels beating between his fingers; in another, he explains how dogs can be conditioned through beatings to hate the sound of the violin (as if this were not also dependent on the player).

René Descartes (1596–1650) justified his casual cruelty in his central work. Animals, he writes, are not only stupid – they have no spirit, no soul or anything that can be considered a sentient self. Rather, it is clear that 'they are destitute of reason, and that it is nature which acts in them according to the disposition of their organs: thus it is seen that a clock composed of only wheels and weights can number the hours and measure time more exactly than we with all our skill'.[10]

According to Descartes, holding a living animal's heart out of anatomical curiosity is no different from stepping on a blade of grass. Both feel nothing, only following the diktat of their construction. Animals are things, not sentient beings, even if the stirrings of their organs simulate something resembling an individual. They are good automata, but nothing more.

Descartes marks an interesting point in the development of the domination of nature as an enlightened dogma – namely, the point at which the propagandist leaves behind all evidence to prevent observable reality from damaging the sublimity of the dogma. Considering how much time Descartes spent with his dog, it is hardly plausible to suppose that he did not see any personality, empathy, independent will, memory or emotion in Monsieur Grat. As a philosopher, however, he defended his position that an animal was merely a *res extensa*, an extended thing, in contrast to the sublime *res cogitans*, the thinking matter from which God and the angels were made and of which humans could partake thanks to their soul, which were likewise not composed of extended matter.

This two-substance doctrine carries an echo not only of Gilgamesh, who was two-thirds god and one-third human, but especially also of Plato's world of forms or ideas – the genuine truth that is represented only imperfectly by the world of perceptions.

In the seventeenth century, in the course of a philosophical debate that was increasingly informed by scientific successes and worked more with analytical and empirical arguments, this theory could help to solve one of the greatest philosophical problems of the time: how can the immortal soul be reconciled with natural science? And, lurking behind this question, was a second, too terrible to be confronted directly: how does science go together with the revealed truth of God?

While the worldly power of the church was at its zenith, it had long been on the defensive in theological terms. The Reformation had made deep cracks in its dogma and authority; the Renaissance had celebrated the possibility of a morally thinking culture without Christianity; the discovery of new continents revealed a world that was not mentioned anywhere in the Bible – other cultures and religions of which it knew nothing; the publishing industry broadcast new heresies and unpleasant truths into the world on a daily basis; and more and more scholars pushed their way onto the stage with theories about nature in which God no longer appeared, asking questions to which the catechism had no answers.

With great effort, it was possible to control all of these developments. In the wake of Counter-Reformation, theologians and artists within the Catholic Church worked frantically on a new, more individualistic and impressive representation of faith. Church architecture adopted Renaissance aesthetics, the so-called New World became a missionary territory, books were confiscated and burnt, printers had their fingers broken and scholars could also be intimidated and eliminated, or induced by generous offers to change sides. Heresy could be universally suppressed, countries and souls could be regained, but all of this consumed strength and could not silence the quiet voice of doubt that was whispering ever-new questions into so many contemporaries' ears.

Let us return to Monsieur Grat, whose master, contrary to all evidence, described him as a mere automaton. Descartes's arguments about the two substances, *res extensa* and *res cogitans*, were intended to do away with a theological doubt by philosophical means: firstly, how God's existence can be reconciled with a material world and how the two are connected; and, secondly, how humans can justify treating other creatures as they do.

This potentially surprising moral concern is especially clear from a letter to Descartes from the philosopher Henry More, who taught at Cambridge, in which the latter defends his two substances theory. More turned to him because of a profound moral conflict, for he saw a contradiction between Descartes's imposing system and his own perception:

> Here, the gleaming rapier-edge of your genius arouses in me not so much mistrust as dread when, solicitous as to the fate of living creatures, I recognize in you not only subtle keenness, but also, as it were, the sharp and cruel blade which in one blow, so to speak, dared to despoil of life and sense practically the whole race of animals, metamorphosing them into marble statues and machines.[11]

Descartes was unmoved by so much sentimentality. He could only accept the intelligent use of language as proof of a soul, and animals could not meet this requirement. Strictly speaking, then, it was only impossible to prove that animals have a soul, which does not prove that they have none. Descartes knew that making this concession would weaken his position; absence of proof is not proof of absence, as More may have thought to himself. So Descartes added a second argument to his first: 'And thus my opinion is not so much cruel to wild beasts as favourable to men, whom it absolves, at least those not bound by the superstition of the Pythagoreans, of any suspicion of crime, however often they may eat or kill animals.'[12]

Was this position the result of a rational analysis? Perhaps not exclusively, the Frenchman admitted, but it was good for society: 'Nor do I tarry at the shrewdness and cunning of dogs and foxes, or at any other deeds performed by brutes for food, sex or fear. For I freely avow that I can most easily explain all those things as arising from the sole configuration of the parts of the body.'[13] The French master thinker conceded that the human mind penetrates the hearts of animals, and that all observation 'from earliest childhood' shows that animals built in a similar way to humans also feel similarly, but he considers the opposing arguments even stronger – otherwise, 'worms and gnats' would also have a soul.

In this letter, Descartes argues in abbreviations, as it were: a conversation between colleagues. Why does he not tarry at the shrewdness of animals? Because there are all manner of arguments about the inner life of animals, most of them motivated by very transparent interests: our eating habits and personal enjoyment make it inconvenient to grant animals a soul. Hunger and the fear of an untamed nature make people cruel towards animals. Humans attribute qualities to animals that mirror human interests and behaviours. The uniformity of general thinking lends legitimacy to these prejudices. It would cause too much of a muddle if philosophers turned their attention towards the shrewdness of wild foxes and tame dogs, the maternal feelings of cows and pigs or a horse's fear of the slaughterhouse. Some things are better not examined too closely.

Perhaps a certain vanity was also involved in this defence, since Descartes did not want the enticing purity of his theory to be spoilt. He was proud of having accomplished a truly historic achievement, a stroke of genius worthy of a Plato or Aristotle. His system, he confidently claimed, had laid a rational foundation for the material world and the existence of God, without obscure myths or clouds of incense. He had proved through reason alone that God exists, that he is good, that he created the world, that humans perceive the world as it is and can decipher its truth through reason, and that only humans are equipped with a soul that connects them to God. He had placed the world on a new footing through reason – through his own reason – so Monsieur Grat and Henry More could scratch themselves all they liked.

Not all philosophers shared Descartes's view of himself or his proof of God's existence. It was not only circular, his colleagues argued, but had also been refuted since the Middle Ages, and they also saw glaring omissions in his epistemology. Despite these scholarly counterarguments,

Descartes's theory proved immensely influential in history – not least because it made peace between theology and the newly invading sciences. From now on, theology could devote itself to the immaterial souls, and science could make do with extended matter: a clear division of powers that at least allowed theology to occupy a certain high ground as it slowly retreated from explaining the world.

One can sense a certain aggression in Descartes's well-formulated Latin sentences, a reaching for power. Nature as a dead, mechanical object finally seemed ready to be conquered and appropriated by the intellect. Two generations earlier, another Frenchman, Michel de Montaigne (1533–92), had thought very differently about his pet. What characterizes Montaigne as a philosophical adventurer is his willingness to follow his observations and their logic to their logical conclusion and develop them further, especially when the results disconcert and surprise him. In the case of his cat, this led him to a famous question: 'When I play with my cat, who knows if I am not a pastime to her more than she is to me?'

Time and again, Montaigne writes, humans try to draw lines between humans and animals, but 'Presumption is our natural and original malady. The most vulnerable and frail of all creatures is man [...] and in his imagination he goes planting himself above the circle of the moon, and bringing the sky down beneath his feet.'[14]

It is humans who ascribe properties to themselves, Montaigne writes, divine properties that set them apart from the other creatures and give them authority over them, since they consider them stupid and devoid of will:

> How does he know, by the force of his intelligence, the secret internal stirrings of animals? By what comparison between them and us does he infer the stupidity that he attributes to them? [...] It is a matter of guesswork whose fault it is that we do not understand one another; for we do not understand them any more than they do us. By this same reasoning they may consider us beasts, as we consider them.[15]

Merely observing animals on a farm can teach a person that they are intelligent, display complex emotions and can deceive, threaten, comfort and understand one another. Their skills and organization by far surpass those of humans: 'Is there a society regulated with more order, diversified into more charges and functions, and more consistently maintained, than that of the honeybees?'[16] Instead of admiring their relatives and learning from them, however, humans consider themselves so superior that, he adds drily, 'their [animals'] brutish stupidity surpasses in all conveniences all that our divine intelligence can do'.[17]

It is the stupidity and greed of humans that literally enslave animals, just as they also abuse other humans and believe themselves superior. The problem with this form of presumption, however, is that it forces all thinking into a logic of subjugation. How else could one explain the fact that humans can willingly enslave themselves, and even go to their death, for the sake of an idea? Ultimately, humans behave no more rationally than animals, for 'We are neither above nor below the rest: all that is under heaven, says the sage, incurs the same law and the same fortune.'[18]

'If Only I Could Paint His Spirit!'

Montaigne's attentive observations and his courage in giving his thoughts free rein, even when they defied the conventions of his time and the dogmas of faith, made him a personal friend to many generations of reading and thinking people who admired, and still admire, this very freedom. For the increasingly clearly formulated project of subduing the Earth, however, this thinking was useless. Nor was it needed, for Montaigne's contemporary Francis Bacon (1561–1626), himself an enthusiastic – albeit not fully convinced, clearly – reader of the famous *Essays*, developed the appropriate theoretical foundations.

Bacon was one of the most colourful figures in British history: a career politician and schemer, observer of witch trials and Lord Chancellor of the Crown, but also a scientist and philosopher, as well as a wonderful stylist whose writings would still be immensely influential long after his death. In later debates, he was often characterized as a textbook example of the mentality of Western exploitation and instrumental reason, but the truth, as so often, is more complex and interesting than the emoji.

Time and again, Bacon is quoted with his memorable motto 'knowledge is power', often with the addition that one must put nature 'on the rack' to wrest away its secrets. The first quotation is correct, though it sounds different in context. The second is a complete fabrication. Bacon did, however, write in *The New Organon*: 'For man is nature's agent and interpreter; he does and understands only as much as he has observed of the order of Nature in work or by inference; he does not know and cannot do more.'[19]

Already as a child, Bacon, the son of a politically well-connected and educated family, was considered exceptionally brilliant. Sickly from an early age, he was educated at home before beginning his studies in Cambridge at the age of twelve – though at that time, when the university was essentially an institution for looking after adolescents from affluent families, this was not necessarily based on intellectual achievements. But Bacon was born for the life of a scientist, many decades before science (partly thanks to him) was invented as a method and academic discipline. When he was eighteen, he was painted by the famous miniaturist Nicholas Hilliard; with unshakable self-confidence, the young man looks the painter directly in the eye. His mouth is closed, his eyes alert and a little arrogant, and his hair is slightly wild, as befits a young man – especially one whose neck is enclosed by a white ruff. This is a resolute spirit who will go far. His head is surrounded by an inscription in golden letters: *Si tabula daretur digna animum mallem* – 'If only I could have painted his spirit.' It is unknown whether the words came from the painter or the painted.

Bacon's career as a lawyer, parliamentarian, courtier and professional politician is the stuff of historical drama. It included murder plots, spectacular trials and the highest political offices, and culminated in his conviction for corruption and withdrawal from political life. This finally gave him the time to return to his scientific and literary interests.

Bacon's writings show him as a man standing between two intellectual cultures and two times. As a boy he had been educated at Cambridge according to the medieval curriculum, and as an adult he participated in witch trials. But he was also one of the most far-sighted critics of his own time, and formulated principles of scientific thought and action that were criticized and further reflected on over centuries, but never replaced.

As a jurist and politician, Bacon thought productively in conversation or correspondences with a variety of people, including the Italian thinker Bernardino Telesio (1509–88), who had proposed a revolutionary theory of nature in his work *De rerum natura iuxta propria principia* (On the Nature of Things according to Their Own Principles, 1565); it is no coincidence that its title recalls *De rerum natura*, the masterpiece by the Roman poet Lucretius. Telesio claimed that the natural world could not be understood based on the Bible or the theories of Aristotle, but that all events in nature are comprehensible only by their own logic and an unprejudiced observation thereof. The world had suffered enough because of unfounded philosophical speculations.

This observation, the Italian wrote, had convinced him that only two principles are effective in nature: heat and matter. Active, living heat came from the sun and matter from the cold, lifeless Earth, and all change was a result of expansion and contraction. Heating and cooling, agitation and calming. This applies not only to living beings, for all things in nature sense whether they need cold or heat to develop; all parts of nature are capable of perceptions and act according to their own specific laws. Understanding these laws means decoding nature itself. This nature, however, reveals itself as a form of living being that breathes, grows and dies in the interplay of heat and cold, Earth and sun, and keeps on adapting and changing based on its own laws.

Telesio's somewhat simplistic theory of nature between heat and matter clearly borrows from the classical two-substance theory and the theological idea of a cold, passive Earth. In this regard, it is not particularly original. His method, however, was both daring and revolutionary: it explained nature purely with reference to itself, without any reference to God. Nature itself was more than an empty space waiting to be filled with meaning and activity by humans; it was an organism infused with a sort of consciousness, full of creative change, and obeyed only itself. The Italian also took man down from his pedestal, again because of his (very male) view:

> We see that humans are formed from the same things as other animals, and that they have the same abilities and organs of nutrition and reproduction, and that they produce very similar sperm and ejaculate it in the same way and with the same pleasure [...] and afterwards grow tired, and that the same things are formed in both cases from the semen, namely the same system of nerves and membranes. And it is the same spirit whereby all animate beings perceive and are able to move in the same way, based on the same predisposition.[20]

We can scarcely imagine today what a bombshell this would have been for contemporaries reading those lines. What they meant was no less than the end of the world – or at least one world. The new order dreamt of by Telesio did not contain any God, although the philosopher half-heartedly added him towards the end of the work, as if inviting an unloved relative to a family celebration. Such a gesture was immediately seen for what it was: the author's way of staying out of prison. At that time, informed readers always read between the lines. This material world had no real place for God, myths, miracles or occult forces, even though the world itself was a miracle that humans, as a small part of it and one animal species among many, had not remotely understood or fathomed.

Francis Bacon too was fascinated by the intellectual audacity of the Italian scholar, who described the Earth like an organism that created and developed itself of its own accord. Those with knowledge of nature – 'Mechanic, mathematician, physician, alchemist and magician' – worked 'to little effect and with slender success', for, as Bacon wrote, very much in the spirit of Telesios: 'The subtlety of nature far surpasses the subtlety of sense and intellect, so that men's fine meditations, speculations and endless discussions are quite insane, except that there is no one who notices.'[21]

The situation of genuine, reliable knowledge about the world was dramatic, the Lord Chancellor wrote, for even in the great libraries, an inquisitive visitor will observe that 'there is no end to repetitions, and how men keep on doing and saying the same things', and 'will pass from admiration of variety to amazement at the poverty and paucity of the things which until now have held and occupied the minds of men'.[22]

The problem with research was easy to understand: too much speculative philosophy, too much copying of old authorities and too little experimental knowledge. Language, with all its imprecisions, was also an obstacle to perception and made it impossible to see things without prejudice: 'The syllogism consists of propositions, propositions consist of words, and words are counters for notions. Hence if the notions themselves (this is the basis of the matter) are confused and abstracted from things without care, there is nothing sound in what is built on them. The only hope is true induction.'[23]

Bacon developed an early sociology of culture when he wrote about illusions such as the tribe, the cave, the market and the theatre, which distort human thought:

> The illusions and false notions which have got a hold on men's intellects in the past and are now profoundly rooted in them, not only block their minds so that it is difficult for truth to gain access, but even when access has been granted and allowed, they will once again, in the very renewal of the sciences, offer resistance and do mischief unless men are forewarned and arm themselves against them as much as possible.[24]

These illusions of the mind come close to what is described here as a collective experience or collective delusion: a shared and practically implemented notion of the world and our place in it which becomes so strong that there is an increasing tendency to tailor reality to the story. This is partly because human perception itself is not objective: 'The assertion that the human senses are the measure of things is false; to the contrary, all perceptions, both of sense and mind, are relative to man, not to the universe. The human understanding is like an uneven mirror receiving rays from things and merging its own nature with the nature of things, which thus distorts and corrupts it.'[25]

What psychologists today call cognitive distortions, which have become a Nobel Prize-worthy field of research, were already described with great precision by Bacon:

> Once a man's understanding has settled on something (either because it is an accepted belief or because it pleases him), it draws everything else also to support and agree with it. And if it encounters a larger number of more powerful countervailing examples, it either fails to notice them, or disregards them, or makes fine distinctions to dismiss and reject them, and all this with much dangerous prejudice, to preserve the authority of its first conceptions.[26]

Bacon's seismographic intellect registered that a new age was dawning in which many things would fundamentally change, but that his contemporaries had great difficulties imagining something truly new that did not 'flow through the usual channels'.

If, before the invention of gunpowder, some resourceful mind had described a cannon based only on its effects – namely, as a weapon that can destroy walls over great distances – then experts, based on their experience, would have imagined longer battering rams and stronger catapults, which would have led nowhere.

> But a fiery wind so suddenly and violently expanding and exploding would have been unlikely to occur to anyone's imagination or fancy; since he would not have seen an example of these at first hand, except perhaps in an earthquake or thunderbolt, which men would immediately have been likely to reject as monstrous forces of nature not imitable by human beings.[27]

The subtlety of Bacon's thought on nature cannot be reduced to the motto *scientia potentia est* [knowledge is power], even though he coined it. He understood that he was living in a period of radical change, and that the mental equipment of his contemporaries was not up to it. Humans are sluggish in their thoughts, cling to absurd traditions and work with false, terribly crude concepts. Bacon's own ambition went further: he did not want to be a mere servant of nature, but to learn to control it by – just like Telesio – understanding it from the inside.

Although Bacon never demanded that nature be put on the rack to wrest away its secrets, he did come close when he postulated that the interpretation of nature should proceed from an *inquisitio legitima*, a legitimate inquisition. As an observer and judge in trials of defendants accused of witchcraft, secret Catholicism and high treason, the jurist Bacon knew exactly what he was talking about. This process was clearly antagonistic, characterized by coercion and the threat of torment. The spirit had to 'penetrate nature', Bacon wrote, in order to understand its automaton-like operations.

Unlike Telesio, Bacon did not see nature as a quasi-conscious organism whose individual elements (including humans) were constantly interacting sentiently. To be sure, Bacon's nature was also dynamic and full of secret mechanisms, such as 'conflict between the different species'[28] and the 'transformation of concrete bodies from one thing into another'[29] – so he was already imagining a kind of evolution – but his writings do show a considerable conceptual distance between humans and their objects, whose secrets will be understood through determination, and by force if necessary.

Nonetheless, it is remarkable how widely European authors around 1600 still differed in their thoughts about nature. Telesio and Montaigne described nature with staggering self-evidence as a web of changes, mutual dependence, various equally valid perspectives and even different forms of consciousness, in which humans wrongly arrogated a special place for themselves. Francis Bacon called himself an agent of nature, in the same way that he may have considered himself an agent of the law – that is, as an active investigator of unsolved questions whose admiration for his object did not prevent him from resorting to brutal methods. Descartes finally admitted in private letters that his image of nature also supported the opinions and interests of the majority, but he defended it in his books to the last drop of ink: only man has a soul, the rest of nature consists of insensate automata – matter whose purpose is to help man fulfil his divine mission by means of his reason, which means dominating them. *Down, Monsieur Grat!*

The Canon and the Antichrist

Within this range of perspectives (there were others too), all four authors continued to be read throughout the sixteenth century, with frequent reprints and translations into several languages. Two of them – Montaigne and Telesio – came to be seen as literary figures or scientific mavericks, and were consulted mostly by aficionados and historians. But the other two became central to the debate, influenced generations of scientists and philosophers and were accordingly celebrated as intellectual founding fathers of a new era, and their writings were admitted to the canon of important works and taught at schools and universities.

Until the nineteenth century, this canon had a lacuna, a glaring omission in the history of thought, a name from the subsequent generation that could not be uttered – or, if so, then only to excommunicate him: Baruch de Spinoza (1632–77). The work of the Dutch-Jewish author and lens grinder was long considered too scandalous, too subversive, to be quoted and discussed openly, and while his admirers celebrated him as a kind of philosophical messiah, his many enemies accused him of casting innocent souls suddenly into the flaming jaws of atheism. Some commentators even portrayed him as the Antichrist himself.

There is a certain irony to this accusation, for Spinoza claimed that he was merely proving God's existence, omnipotence, goodness and perfection with mathematical precision, and thus formulating valid principles for a life based on divine reason using strict logical criteria. The devil, as always, was in the details.

Spinoza had grown up in a Portuguese-Jewish merchant family in Amsterdam and attended a yeshiva, where the students were instructed in the Talmud and biblical exegesis. But his lively mind craved to explore broader horizons. He learnt Latin, began moving in the circles of free spirits and dissidents and was finally expelled from the Jewish community for his heretical views. In addition, they placed him under a curse: from that day forwards, no member of the community or his family was allowed to speak to him or have any contact with him.

Spinoza left his home town and made a living constructing optical instruments and grinding lenses. According to the reports of his few but devoted friends, he lived simply, almost like a monk, working on philosophical texts. As one of the rare participants in debates in a Europe whose geographical and intellectual perspectives were radically expanding, Baruch de Spinoza was versed in the discursive methodology of both the Jewish and the Latin-Christian traditions. This allowed him to view each from the perspective of the other and question it argumentatively from an unusual position.

To speak of God, Spinoza first had to speak of nature, for God was ubiquitous, perfect and almighty; the world was unthinkable without him, he was the groundless ground and the substance of all that existed. Thus, God is in everything and nothing exists outside of him. Up to this point, his logic would have convinced the most orthodox theologian or censor, but it led to a dangerous converse argument: if God is in everything, if his perfect will created the world as it is and cannot change it, then he himself is not only immanent in all things, but is no longer necessary as an idea. God is matter and the laws of nature; in Spinoza's own legendary formulation, the world consists of *deus sive natura* – God or nature, two interchangeable words.

There had been trouble brewing for a long time in theology, but the following steps brought everything crashing to the ground amid the dust of centuries. God is perfect and good, and the world is divine substance; this means that the world is perfect and good, and follows its immutable laws without pursuing any purpose or goal except to obey its own order. But this also means that neither history nor human life nor nature itself points beyond itself. This, Spinoza writes with the resignation of experience, is too much for many minds. They prefer to entrench themselves in convenient prejudices that make life bearable for them – even if they are based on deception, for example: 'human beings commonly suppose that, like themselves, all natural things act for a purpose. In fact they take it as certain that God directs all things for some specific purpose. For they say that God made all things for the sake of man, and that he made man to worship him.'[30] Humans stubbornly project their own ambitions and their own social order into heaven, as

Spinoza shows with a few lines. It is possible that these lines express personal pain, but it is clear that they reflect the painful recognition of the human flaw of repeatedly choosing the convenient illusion over the inconvenient truth. Rarely has religious criticism been formulated so clearly in so few words.

Humans live without knowing the true causes of things, believing that they are free, 'and never think, even in their dreams, about the causes which dispose them to want and to will, because they are ignorant of them.'[31] Humans act purposively to satisfy their urges, and therefore see only purpose and use in nature. Because they themselves were born into that nature, however, and experience it as something outside themselves, they assume that someone else, namely God, must have created nature for their use, as a spiritual bargaining chip, for humans must worship their gods in return:

> This is how it came about that they each invented different ways of worshipping God based on their own character so that God would love them more than other people and direct the whole of nature to the service of their blind desire and insatiable avarice. This is how the prejudice turned into a superstition and put down deep roots in their minds, and this is the reason why they have each made the most strenuous endeavour to understand and explain the final causes of all things.
>
> But in striving to prove that nature never acts in vain (i.e. not for the use of human beings), they seem to have proved only that nature and the Gods are deluded human beings.[32]

The mild-mannered scholar did not write a word of criticism about the creator (or nature), but he certainly attacked human stupidity, which, in its narcissism, could see nothing but mirrors around it. Naturally, it was an illusion to think that the world had been created for human use, and they could bribe a God to have their way, but humans themselves preferred this illusion to the truth, since they identified with it and felt good about it, and because their neighbours thought the same. It was simply too great an effort for them to 'overthrow the whole structure and think out a new one'.[33]

This criticism struck the debates about the one, indivisible truth of God and his blessings on the weapons of various armies, about the God of slaveholders and colonial overlords, of science and the poor, to the core. Theologians and philosophers realized that this argumentation was a slippery slope and would inevitably lead to their ruin, and declared Spinoza's thought 'the spawn of hell, written by the devil himself'.

In the context of our story, however, Spinoza's understanding of nature is more important – and at least as revolutionary as his concept of God. As an attentive reader of Montaigne and Bacon as well as Telesio and Descartes, he was familiar with his predecessors' models of nature and constructed an argument of unparalleled elegance, as if Montaigne had written it for Descartes. Nature is an infinitely complex system whose laws are ignored and twisted by humans out of ignorance and greed. This results not only in a degradation of nature, which is viewed merely as a means to an end, but also to confusion among humans themselves, who subject themselves to the same logic, pursuing unrealizable goals rather than analysing causes and effects and using the understanding of inescapable causality thus gleaned as the foundation for their own happiness and virtue.

Spinoza demolished too much of the edifice of theology to build it up again. His critics feared that only ruins would remain once his works were read, and those people sat in high places. As a result, his books were placed on the index of forbidden books by the church and banned in several countries, but they were also sold at high prices as manuscripts, pirate editions and translations, or passed on from hand to hand.[34]

This secret dissemination created and nurtured a small but very lively current of thinkers searching for perspectives beyond the logic of masters and slaves. We will encounter some of these later. But their efforts were barely visible amid the general move towards the new gospel of the scientific and rational domination of nature, which became the motor for prophets who all believed they were striving towards a New Jerusalem, be it heavenly or not so heavenly, even when they could hear the bones crunching beneath their feet.

*

Descartes's statement that humans are 'masters and possessors of nature'[35] was repeated in countless modulations by theological and enlightened authors alike. The most impressive evidence of this mentality, however, can be found between tall hedges, not book covers. The park of Versailles (here depicted in a copperplate engraving from 1661 – see figure 9) was a perfect implementation of Descartes's philosophy of nature: everything that is visible in it exists to express, to symbolize, to celebrate and to stage the king's power. In the centre stands the god Apollo, whom Louis XIV, an excellent dancer, liked to play on the palace stage during ballet evenings.

The visual axis is strictly vertical and shows, behind the Fountain of Apollo, the large reservoir, where one sees not only Venetian gondolas (which could be steered by gondoliers imported from Venice) but also several warships, which could entertain the court with staged battles. The trees around them are arranged in rank and file like an army of leaves and trunks, ever ready to pay homage to the monarch. No natural expression of nature is allowed. Everything is trimmed, demarcated, watered via pipes and canals and raked, planted, plucked and uprooted by an army of gardeners. The trees adorning the park were dug up somewhere else in the country and brought to Versailles (tens of thousands died on the way), the plants were bred in hothouses, and the flowers were planted in geometric patterns just when they were in bloom, for the rational

9 Adam Perelle, *The Fountain of Apollo* (Versailles), copperplate engraving, 1661

133

order of the human mind had to rule every detail. The monarch's power extends to the horizon – but no farther. Even absolute glory has a visible limit.

Paradoxically, this propagandist intensity showed that, by the seventeenth century, the idea of a king who received his power directly from God had long been in need of justification and demanded extravagant performances like the lion hunts of the neo-Assyrian kings, who likewise invested immense resources in the representation of their power and possibly had to do so.

The parallel between the architecture of European parks and the Mesopotamian rulers is no fanciful one. There, too, gardens and parks were highly valued as demonstrations of a ruler's power, from the Hanging Gardens of Babylon (perhaps the famous garden of Ashurbanipal in Nineveh) to the garden in which Inanna awaited her lovers under a tree, and the idea of the *pardes*, the civilized space that kept the wilderness at a distance by means of a wall. This played an important part in Mesopotamian culture and is the origin of the biblical paradise.

Parks and gardens defined inside and outside; they dramatized the gulf, as well as the connections, between nature and culture. The *hortus conclusus* of the medieval monasteries and nobles was a place in which the allegorical order of creation was realized, the will of the creator himself, while the wilderness and its dangers threatened from without.

Something else was also present whenever power and subjugation were staged: a voice that whispered quietly but unmistakably, against the chatter of courtiers and the blaring fanfares, a series of warnings that did not fall silent. *You are still king here for now, but the limit of your power is over there, the line of treetops on the horizon: that is the forest, the wilderness, the end of the performance. The forest negates eternity, and if you chop it down, and then another one, eternity will still always end – at a forest, where the wilderness begins. Memento mori.*

An Experiment on a Bird in an Air Pump

10 Joseph Wright of Derby, *An Experiment on a Bird in an Air Pump*, 1768, oil on canvas, 183 × 244 cm. National Gallery, London

The experimenter looks directly at me. Long hair falls across his haggard face; he seems to be challenging me to react somehow. Everything depends on me, the viewer; I have to decide quickly. The bird in the glass has only a few moments left to live if no one speaks the saving word and lets the air pumped out of the glass flow back into it.

But who will speak it? Those present are strangely disinterested. The two young men at the bottom left watch attentively as the animal fights for its life. The young lovers behind them – the gentleman wearing a strikingly patterned waistcoat, the lady an ermine collar – have eyes only for each other. The man to their right is comforting a young woman, perhaps his daughter, who is covering her eyes with her hand to avoid the sight; there is no escape from the scraping of the wings and the scratching of the claws against the glass. The man beside her stares ahead mutely as if meditating, while another looks at the clock as if to calculate how long the death throes will last. A boy in the background carries an aviary on a long hook; it is unclear whether he is taking it down to replace the healthy bird after the experiment, or putting the cage away because it will no longer be needed. Only the little girl in the middle looks up fearfully at the animal in its agony. But who will speak the saving word? Who will say 'Enough!'?

At first glance, it is the drama of the moment that draws the viewer so irresistibly into this picture, then it is the crossing of gazes and stories; the hidden messages only come in a third step. But one thing at a time: let us start at the beginning.

Joseph Wright (1734–97), who added his birthplace for the more sonorous name 'Joseph Wright of Derby', was a respected portrait artist who spent the most fertile years of his career in the north of England, where he associated with, among others, the ceramics pioneer Josiah Wedgwood; the textile entrepreneur and inventor of the pneumatic loom, Richard Arkwright; the chemist and theologian Joseph Priestley; and the doctor Erasmus Darwin, grandfather of Charles Darwin. The painter's friends and patrons included some of the leading figures of both the incipient Industrial Revolution and the social movements of his time. Wedgwood not only produced classicist teapots for the middle class; he was also an important member of the anti-slavery movement. Priestley, too, who managed to be simultaneously a materialist philosopher and a vicar with the Protestant Dissenters, had never shied away from controversy. In 1791, a mob burnt down his house and laboratory because he had announced a festive dinner on the occasion of the French Revolution.

Wright was at the centre of a newly emerging world in which uncanny thoughts were being thought, a world whose horizons seemed unlimited; here was the engine room of a new economy, a new society, a new science and a new world order. The future was being designed in the northern England of the 1760s. Josiah Wedgwood owed his fortune to a methodical, years-long search for new ceramics production processes, and the textile baron Richard Arkwright stood like no other for the future of the industrialized world. He was an energetic and hot-tempered self-made man of humble origins who had doggedly worked his way up from a hairdresser and wig-maker to a mechanical engineer and inventor. Eventually he owned several factories, and two-thirds of the workers were children on thirteen-hour shifts, since their little hands were better suited to the looms.

No one got anything for free in this revolution. The cotton that was processed came from India, where producers were threatened with severe punishments if they spun and wove their own cotton. Instead, they were forced to sell the raw material to their English colonial masters and re-import the finished product at a much higher price. Such measures turned India, which had been the second-largest economic zone around 1700, into one of the poorhouses of Asia, while new factories sprang up like mushrooms in northern England. England's industrial miracle was not only based on English diligence.

The painter was more than merely an illustrator of this milieu. He keenly and attentively registered the ambitions and dreams that carried the protagonists of this new reality. In his large-scale painting, Wright depicts an experimenter at the dramatic climax of his experiment. The vacuum had been a known phenomenon for over a century, but vacuum pumps had long been unaffordable and those interested in science only knew them from books. Now it was finally accessible in a domestic setting: fun for all the family. Because one cannot see a vacuum, its effect was illustrated using living creatures. The white parrot in the glass vessel is already at the end of its tether. At the beginning of the experiment, when the air was sucked out of its prison, it will have flapped in panic, but now it no longer has the strength. The vacuum takes effect; the scientific demonstration has succeeded. Show experiments of this kind were so popular that admittance was charged for public presentations. Sometimes the test animal was snatched from the jaws of death at the last minute.

The strong contrast between the brightly illuminated faces and the blackness around them directs our attention away from the science and towards their emotions. Between obliviousness and amorousness, their own pain and complete indifference, the members of the group seem mostly concerned with themselves. Their diverse emotions testify to an unprecedented emphasis on individuality and state of mind, but they remain unconnected to the situation at hand: the suffering of a living creature in fear for its life.

The experimenter is the only one seeking contact, his gaze turned outwards. He, too, ignores the bird. He seems tense, waiting – perhaps for the saving word. But will that salvation come? The white-haired experimenter, the man next to him, whose hand points upwards like that of a Baroque Christ, and the bird itself form a triangle, an ironic trinity of Father, Son and Holy Parrot.

Was it Joseph Wright's intention to celebrate the curiosity and entrepreneurial spirit of his contemporaries, or does his picture express a critical view of their infatuation with progress, which was based on contempt for humans and nature?

The idea of dominating nature was especially successful in such societies as France, the Netherlands and Great Britain, which actually experienced a certain self-empowerment during the seventeenth and eighteenth centuries. Not only the growing colonial empires and booming mass production of consumer goods such as textiles and ceramics, but also the spectacular successes of their scientific experiments and demonstrations, as well as the increasingly confident demeanour of a new bourgeois culture, gave the impression that a true change had begun, that reason would finally begin its triumph and crush hunger, disease, war and all the ills of this world beneath its noble feet. Finally, what had been interpreted for centuries at best as an allegory for human self-control and the suppression of desire now seemed within reach.

New technologies gave humans greater access to nature, enabling them for the first time to expose and manipulate its secret laws for the benefit of humanity. The vacuum pump could make invisible air disappear and take an animal's life, while Signor Galvani's experiment, in which a frog's legs were made to twitch at will using an electric current, touched on the secret of life itself. It was only a short way from here to Mary Shelley's *Frankenstein*.

The public's curiosity about scientific discoveries and the adventures of the new were immense. Newspapers and pamphlets, books and public lectures provided reports of electronic phenomena and daring flights in hot air balloons, disseminated theories and instructions for experimentation mixed with other popular stories about witches and mysterious murders. Scholarly societies sprouted up, and gentlemen put the money they gained from their investments in colonial undertakings into scientific devices and investigations of their own; there was still so much to discover.

Joseph Wright's friend Joseph Priestley alone (who was never particularly well off and always had to fulfil his duties as a vicar, writing hundreds of sermons as well as essays and books on theological subjects) was simultaneously – on the side, as it were – not only the author of an influential work about electricity, but also the first to discover oxygen and its function for the blood system. There was no need yet for great laboratories or special equipment to find a place in scientific history; suddenly, it was conceivable to understand, change and tame nature purely through reason.

Never before had the subjugation of nature seemed so within reach. In addition to these wonderful discoveries, which cast a completely new light on nature and appeared to reveal its secret mechanisms, the European middle class also enjoyed the availability of consumer goods including sugar from the West Indies (grown by slaves); cheap cotton from the northern English mills, brightly coloured with new pigments; porcelain and opium from China; tobacco, tea, coffee and spices from legendary island kingdoms; mahogany furniture, walking sticks made of ivory and ebony and other luxury objects that allowed the world's riches to find their way directly into the hands of the bourgeoisie: everyday symbols of the West's dominance over the rest of the world.

Borne by a wave of optimism, economic upturn and belief in a better future, visionaries had already set their sights on other goals – like the experimenter in the painting, with a living creature fighting for its existence right beside him. His hand is on the valve, but does not move. He does not spare a glance for the animal's suffering.

Nor is there anyone else who might help this dying parody of the Holy Spirit. Joseph Wright of Derby may have chosen the parrot for his dramatic scene because such birds were still quite expensive in England in the 1760s – possibly a comforting indication that such a precious possession will not be killed – but this calculation still does not absolve the observers from their moral failure. Only a child shows pity, while the others present either register the agony objectively or are too self-absorbed to help the animal. But they too know that they should not follow their feminine, emotional feelings.

Perhaps that is what the man is telling the older girl, pointing out the bird's usefulness. It is no coincidence that two girls are reacting emotionally, while the boys stare in fascination at the glass sphere with its suffocating prisoner. Feelings are feminine and reason is male, as is cruelty, or the fine art of emotional dissociation – the second step after the hypocrisy of divinely ordained domination. Progress demands sacrifices, and suffering serves a higher purpose, so there is no place for emotion; one's gaze is always directed at the horizon.

This image of the Enlightenment – rationalist, pitiless, abstract and cruel – fits the caricature mobilized by its opponents against the new way of thinking. They had quite different adjectives in their arsenal, however: blasphemous, presumptuous, dangerous, immoral or diabolical.

In fact, the Enlightenment did not match either its caricature or the secular icon, for, although it triggered an incredible boost of freedom, its own thought structures were not always as revolutionary as its followers proclaimed and its opponents feared.

'The Enlightenment' was never a school of thought with fixed dogmas, aside from an emphasis on human reason, a basic optimism and a certain egalitarian tendency, though the latter had highly disparate manifestations. National debates would often focus on very different issues, and their forms varied widely depending on external circumstances like censorship and repression. Although people were aware of what was going on beyond language and territorial boundaries, the debates in southern Italy concerned quite different matters from those in Scotland, London or Paris, to say nothing of the German provinces or the court of the Russian tsar. Some scholars maintained long correspondences with one another; others travelled to visit important salons, such as that of Baron d'Holbach in Paris – however, despite these multifarious connections, the key issues diverged greatly in different countries and languages.

In addition, not only did the English Enlightenment have different interests from the *Aufklärung*, and the thinkers of the *Lumières* different ones from those of the *Illuminismo* or the *Illuminación*, but their debates also developed an enormous range – from a liberal, rational conservatism with scientific interests to the wildest proto-communism, from theological apologia to the most radical materialism, and from high moral ideals to sheer nihilism. In all these debates, the more intellectually daring and dangerous positions were stymied by the fact that the authors were often either unable to distribute their works or subject to great risks and difficulties if they did. Nonetheless, over several generations, a multifaceted and complex geography of thoughts and conversations grew that cannot be put in a single category.

The rationalist, moderate Enlightenment of Kant or Voltaire, of Hobbes or Leibniz, was perceived by its numerous opponents as an attack on the traditional world order. In fact, however, it also had the opposite function, since it gave many core ideas from the Christian theological tradition a new life in a secular world.

This is an important, often overlooked, aspect of the history and historical effect of the Enlightenment. It not only swept away old structures, but also enabled them to have a new life in a previously unknown and thus scarcely recognizable form. Although the Enlightenment officially attacked a powerful line of intellectual tradition, it simultaneously continued some of its important structures and charged them with new energy. Subjugation, the urge to subdue the Earth, found a secular costume there.

The altars of reason and the cult of the highest being that would bring nostalgic Catholics and effusive enlighteners together in the later years of the French Revolution were merely the caricature of a theological strain of Enlightenment thought from which only few authors truly managed to break away.

What were far more important were the subtler ideas, the basic assumptions on which their ethical, epistemological and historical positions rested. Few authors disputed the position of humans outside and above nature or their role as its conquerors; most of them saw history as a process of progress with interruptions and setbacks, a striving for perfection and freedom that could succeed by privileging reason.

It is remarkable how theologically charged these concepts are, how strongly they adopt core theological ideas – from progress (of salvation history) to the position of humans outside and above nature and the subjugation of the latter (see Genesis) and salvation through reason, which had been the soul in the religious context. These motifs remained stable in European thought, which was possible because Enlightenment philosophers gave them new labels that made people forget their theological origins. In the vocabulary of the Enlightenment, they sounded like rational, even scientific ideas that could be proved by scholars drawing on endless historical parallels and literary quotations.

Adapting the structures of Enlightenment thought to a Christian agenda was a logical choice; most of those thinkers had been brought up as Christians, and those ideas were so deeply familiar to them and their societies that they seemed like the only possible structures in which to think. Although enlightened authors also attacked Christian dogmas, they used arguments and figures taken from the Christian tradition and developed them in their own way.

However, because the moderate Enlightenment had developed its own reinterpretations of theological concepts like the exceptional role of humans, historical progress and the saving power of reason, and thus continued them, it seemed to be following on from the theological and philosophical efforts to justify violent subjugation in the name of the Lord.

The justification industry, originally born of the necessity to reconcile the brutality of a society with the uncompromising message of Jesus, also made use of the Enlightenment as a timely social and philosophical movement whose arguments proved immensely useful. From now on, the busiest justifiers worked on the side of reason and science. Over the centuries, their expertise was used to buttress very different, often contra-dictory positions with theories and data, and the driving forces behind these claims often had more to do with the historical context than with scientific necessity. The arguments varied, but the basic assumptions remained the same – and with them, the paths of power. Whether women were cursed by the legacy of Eve or rather, in the vocabulary of science, to be understood as an emotional, hysterical, inferior version of man, the conclusion is the same: power and decision-making authority are safer with men.

Regardless of whether humans of other skin colours or religions have no souls and are damned to eternal torment in hell, or were instead constructed as 'closer to nature' or 'closer to apes than to civilized man' based on other criteria, the consequence can only be their subju-gation – for their own salvation, as the justification industry among the theologians of late antiquity had ensured that every form of violence could be reframed as God's will and human duty.

The Enlightenment transferred these power structures and their justi-fication into the new vocabulary of science and empirical knowledge. In this respect, it was also a deeply conservative project: it gave a new lease of life to religious concepts that were in the process of being discredited as tools of knowledge and removed from the philosophical debate, and thus secured a place for them in the social discussion to this day.

Although the idea of progress has created increasing doubts, there are still many people in the West who claim that history has a direction and a goal, a telos, and can therefore be temporarily derailed at most. Such veritably theological notions were further developed by a dominant group of Enlightenment thinkers in a new guise as scientific, reasonable ideas.

Humans outside and above nature, destined to rule over it: many Enlightenment thinkers whom the historian Jonathan Israel groups among the 'moderate mainstream' could identify with this profoundly biblical, theological construction, which defied all perception and experience. And not only that: this view of the world inspired and drove them to produce an entire flood of scholarly works, but also to engage in political activism as well as scientific and social experiments.

The Enlightenment's moderate mainstream was characterized by its compromises. It inveighed against superstition, but – like Voltaire – considered religion useful for keeping the masses under control. It poked fun at the cult of saints and belief in miracles, but insisted that there must be an ingenious clockmaker behind the grand clockwork of the universe. It heroized reason but demonized Spinoza, the most reasonable thinker of all. It sought to apply reason uncompromisingly wherever it was useful, but held back any overly far-reaching consequences – with hard sanctions if necessary.

The jurist, mathematician and philosopher Christian Wolff (1679–1754), who taught at the University of Halle, got a taste of these sanctions when, in 1721, he made the terrible mistake during a public lecture of voicing his opinion that Chinese philosophy and civilization proved that it is still possible to live a moral life without Christianity. Within two years, his Pietist opponents managed to have him banished by the king as a blasphemer and forced to leave the city within forty-eight hours on pain of death.

Wolff's career-damaging fascination with Chinese philosophy, which he studied as intensively as the available translations allowed, marked a point of contact between two important philosophical traditions that was also sought by other philosophers. Since the Jesuit mission at the Chinese imperial court, the Middle Kingdom had been known as a civilization that may have lagged behind the West regarding some scientific discoveries, but that was every bit the equal of Europe when it came to its administration, its wealth, its profound tradition, its powerful industry, its trade and its rich culture. This, of course, raised the question of how people who had not yet partaken of God's grace were capable of such moral and cultural achievements. Few thinkers made the effort to study Chinese texts in detail; for most of them, the precedent was astonishing and significant enough – namely that, as Wolff had claimed, a civilized and even commendable life did not necessarily have to be a Christian one.

The Theology of Fish

The danger of enlightened thinking was that it kept leading to such conclusions. Its arguments developed a pull of their own, and could quickly cast a mind taking its first careless steps into the abyss of materialism, atheism and republicanism. If it was possible to live morally without Christianity and explain nature without referring to God or the Bible, then why was God necessary?

Such thoughts led directly to the abyss, so the 'moderate mainstream' Enlightenment spilt an enormous amount of ink to protect itself from its own conclusions, for, in the eyes of many contemporaries, the task of philosophy was not only to explain the world, but also to justify it in its current structures, which often involved economic dependencies – as with Christian Wolff, who lost his livelihood. Many others lost their entire existence or even their lives; we only know about them from interrogations and court documents, as they were put on trial before they could bring up their positions for discussion.[36]

The most successful branch of the Enlightenment's justification industry was that of 'physico-theology', which became an intellectual fad during the eighteenth century. Nothing less was at stake than justifying the Bible in the light of the newest scientific findings and proving that all these discoveries were already contained in the Bible or intended by it. The most successful authors of this genre lived in considerable affluence.

As early as 1713, William Derham, a friend of Isaac Newton and Edmond Halley, published his book *Physico-Theology*, in which he claimed to recognize the wise hand of God in all natural phenomena. But how could suffering in nature be reconciled with the existence of a benevolent God? The Earl of Shaftesbury, who was extremely influential in England and France (and had no need of such success), argued with reference to Spinoza that nature is always good, even when the divine plan may cause individual suffering: 'if the Ill of one private system be the Good of others; if it makes still to the Good of the general System

[...] then is the Ill of that private System no real Ill in it-self; more than the pain of breeding Teeth is ill, in a System or Body which is so constituted'.[37]

The English parliamentarian and writer Soame Jenyns found an even simpler answer to the problem of suffering and injustice. The universe consisted of a system of subordination, he wrote, and in such a system it is natural that those lower down in the hierarchy are less happy than those who have found a place higher up. Just as animals must endure the moods and cruelties of humans because they stand below them, humans too must bear the misfortune assigned to them:

> If we look downwards, we see innumerable species of inferior Beings, whose happiness and lives are dependent on Man's will; we see him clothed by their spoils, and fed by their miseries and destruction, enslaving some, tormenting others, and murdering millions for his luxury or diversion; is it not therefore analogous and highly probable, that the happiness and life of Man should be equally dependent on the wills of his superiors?[38]

A higher intelligence had structured the world just as it should be: this was also argued by the German polymath Georg Wilhelm Leibniz, who contributed not only the infamous theory of the best of all worlds (for a good and wise God could not have created any other), but also a further reflection. The punishment of sinners in hell must last forever, he argued, since the damned in hell curse God in their pain, and thus continue sinning.

The French scholar Noël-Antoine Pluche (1688–1761), a baker's son, unexpectedly became one of the most successful authors in Europe when he turned his desperate political situation to his advantage. After being driven from his position as a head teacher in Reims by political intrigues, he eked out a living as a private tutor and began writing down his imaginative lessons on all aspects of nature in a series of books, which he then published. The nine-volume *Spectacle of Nature* (1732–50) became a publishing sensation, for Pluche put his finger on the pulse of a time that was reacting to scientific insights and discoveries with both fascination and distrust, for new possibilities were always accompanied by new doubts.

The teacher Pluche was writing for a middle class that wanted the new sciences to be embedded and safeguarded, especially for their children.

All natural phenomena, he writes, have a purpose, a role to perform, and give us cause for moral reflection and deep insight: 'They all have a language that is directed at us, and not only us. Their particular structure tells us something. Their properties have a goal and show us the intention of the creator.'[39] Nature is 'the most learned and perfect of all books to cultivate our reason'.

What follows are nine volumes with lively conversations between fictional characters who speak about nature in all its fascinating details and life forms, all based on current knowledge and adorned with generous, noticeably naturalist illustrations. However, the author has a warning for overly curious children and foolhardy adult readers: 'It is not enough to make the spirit curious by guiding it towards beautiful things. One must also warn it and limit its curiosity, and so we conclude the first part with a short meditation on what is appropriate and on the necessary limits of human reason.'[40]

Pluche was encyclopaedic in his ambition to include, explain and theologically categorize all natural phenomena. Even in those days, however, some scholars preferred to concentrate on specialized areas. The historian Ritchie Robertson describes some of the excesses of this discipline: 'In Germany, physico-theology gave rise to a large number of specialisms praising different aspects of creation, such as ichthyotheology (fish), petinotheology (birds), testaceotheology (snails), melittotheology (bees), chortotheology (grass), brontotheology (thunder) and sismotheology (earthquakes), each of which had at least one book devoted to it. The Netherlands produced also theologies of snow, lightning and grasshoppers.'[41]

The theological investigation of nature was a great popular success, but authors who maintained intellectual honesty were confronted time and again with the exact problems of which Pluche had warned. The teacher Hermann Samuel Reimarus worked and wrote in Hamburg, where, very much in the spirit of the time, he wrote an apologia or letter of defence for the reasonable worshippers of God. Christianity, he wrote, is a practical, moral and reasonable religion that seeks to motivate good deeds; it is not an edifice of absurd and corrupt fantasies, false tales and fantastic theology.[42] This attitude also led Reimarus to view some biblical myths critically. He was clearly someone who was in the habit of thinking carefully about things. It is written in the Bible that the

Israelites – 600,000 men with their families – were able to cross the Red Sea within one night thanks to a miracle. The Hamburg grammar school teacher calculated that at least 4 million people would consequently have been involved, held up by mothers with infants, the weak and the old, stubborn livestock and slow carts laden with the people's entire belongings, all on the stony, slippery seafloor; impossible to manage all that in one night.

Lisbon

On 1 November 1755, All Saints' Day, thousands of believers crowded into the churches of Lisbon to attend Holy Mass. Around 9.40 a.m., the city was shaken by a strong earthquake. Church roofs and entire buildings collapsed, burying countless victims beneath rubble; cracks 5 metres wide opened up in the ground; candles lit in honour of the saints fell to the ground and set houses on fire, and soon entire districts were in flames. Within roughly 30 minutes, the streets had been transformed into an inferno.

The survivors fled from the crumbling houses and headed for the port. To their surprise, they saw that the sea had retreated and exposed several shipwrecks in the harbour basin. Then the tsunami came hurtling towards the city and devoured thousands more victims. It is estimated that between 30,000 and 60,000 people perished in this disaster.

One month later, international newspapers began reporting on the catastrophe. The *Hamburgische unpartheyische Correspondenten* [Hamburg Impartial Correspondents] and the *Berlinische Nachrichten von Staats- und Gelehrten Sachen* [Berlin News of State and Scholarly Matters] were among the first, and over 3,000 articles appeared over the course of the year. What preoccupied the authors more than the events themselves, however, was a question: how could a good, all-knowing, all-powerful and reasonable God let his own faithful die so arbitrarily and cruelly, and on All Saints' Day too?

In admirably imaginative fashion, European thinkers set about explaining the devastating natural event as part of God's plan. Jean-Jacques Rousseau was appalled that humans blamed God in their stupidity, rather than themselves for insisting on flocking together in cities. Pious commentators felt obliged to point out that Lisbon's red light district was directly by the docks, and God clearly wanted to punish the sinners. Others saw the earthquake in an even larger historical context, namely as a punishment for the depravity that had befallen Christianity, and yet others quoted the Englishman Thomas Burnet, who had already written in 1684 that disasters are necessary to remind humans of the 'Emblems and Passages of Hell', and also to relieve the seething of the Earth's interior and thus prevent even greater disasters.

But many of their contemporaries were neither willing nor able to follow them. If little children and pious Christians were being swallowed up by the earth and crushed by church roofs, it was time for some fundamental rethinking. Opinions differed in particular about the role of God in the new reason-centred universe of enlightened thinking – a debate that was never purely theological in nature, but always had political implications too, since power drew its legitimacy from the grace of God.

Immanuel Kant, at the time a young man looking out for a job at the university, took the position in his first extended work, *Universal Natural History and Theory of the Heavens* (1755), that occasional disasters were the price of a creative universe – little more than cosmic work accidents, like the countless insects and flowers that are destroyed by a single night of frost without making nature any poorer:

> The deleterious effects of infected air, earthquakes, floods eradicate whole peoples from the face of the earth, but it does not appear that nature has suffered any disadvantage through this. In a similar way, whole worlds and systems leave the scene after they have finished playing their roles. [...] Meanwhile, so that nature will beautify eternity with changeable scenes, God remains busy in ceaseless creation to make the material for the formation of even greater worlds.[43]

Kant had taken an argumentative gamble. It is easy to infer the existence of a creator from the harmony and beauty of nature, but what about disasters like earthquakes? Do they disprove the idea of a good God, or can they also be integrated into the grand plan? Yes, he answered, citing every conceivable modern and classical authority to support his case. The creation is infinite and its purpose unfathomable from the perspective of small human concerns and needs. Worlds and galaxies form, creation and destruction grow to ever greater perfection. The young scholar, who came from a strict Pietist family, concluded the following from this: 'Let us therefore accustom our eye to these frightening upheavals as being the ordinary ways of providence and even regard them with a kind of appreciation.'[44] The only dissonance in this choir of acrobatic justifications came from Voltaire, whose great 'Poem on the Lisbon Disaster' is a reckoning not only with the apologists, but also with Leibniz's theory that this world is necessarily the best of all worlds, and that all human suffering therefore stems only from the fact that humans cannot understand the true inner workings of the creation – a subject the cynical moralist Voltaire would take up again in his novel *Candide*.

However, Voltaire's angry reaction to the discussion about Lisbon shows no signs of cynicism or his usual amused aloofness. His verses attack bigots of all stripes who have the audacity to proclaim 'Tout est bien':

> To those expiring murmurs of distress,
> To that appalling spectacle of woe,
> Will ye reply: 'You do but illustrate
> The iron laws that chain the will of God'?
> Say ye, o'er that yet quivering mass of flesh:
> 'God is avenged: the wage of sin is death'?[45]

It is notable what a profound personal effect this controversy had on the poet. It inspired him to his darkest moments, which make him seem like a direct precursor to Sartre or Camus:

> Man is a stranger to his own research;
> He knows not whence he comes, nor whither goes.
> Tormented atoms in a bed of mud,
> Devoured by death, a mockery of fate.

In his *Ideas for the Philosophy of the History of Mankind* (1781), Johann Gottfried Herder sums up the debate with a certain detachment. As a Christian poet and preacher, he was himself a man of God, but saw him as a purely ethical authority that does not intervene in the blind course of nature:

> The cry raised by Voltaire at the Lisbon earthquake was most unphilosophical, since his indictment of the Deity on that account verged on the blasphemous. Are we not indebted to the elements for ourselves and everything that is ours,

even our abode, the earth? If the elements, in accordance with ever-acting laws of Nature, periodically rouse themselves and reclaim what is theirs; if fire and water, air and wind, which have made our earth habitable and fruitful, continue in their course and lay waste to it [...] what would happen other than what must happen in accordance with eternal laws of wisdom and order? As soon as a Nature full of mutable things is set going, there must also be a going under; or rather, an apparent going under, an alternation of shapes and forms. But this never affects inner Nature, which, high above all ruins, always rises like a phoenix from the ashes and blooms with youthful forces. The formation of our mansion and all the substances that it could produce must prepare us for the frailty and vicissitudes of all human history. With every closer inspection we recognize this more clearly.[46]

Lisbon became synonymous with the analytical weakness of rational religion. For the educated elite, at least, the earthquake of 1755 became a mindquake. The consequences of these mental tremors extended from the political reforms in Lisbon pushed through by the enlightened Marquis de Pombal to a sense of intellectual security that challenged thinking and reading people everywhere to rethink the relationship between nature and religion. The earthquake in Lisbon polarized the public debate on the relationship between humans and nature, which had grown unstable in the wake of new and spectacular scientific successes.

The majority of authors who wrote about this relationship did so with varying degrees of explicitness in order to reconcile religion and science, even though this required an increasingly difficult theological balancing act. The presence of theological thought within the Enlightenment is obvious here; what was considerably more important, however, was the more profound influence of theology on the enlightened project itself.

Practically all Enlightenment thinkers (with the exception of a few Jewish ones) had been educated at Christian institutions and by priests or monks, and some had undergone intensive theological training, since schools like those of the Jesuits for gifted boys were the only path to higher education. Girls barely had such possibilities, and the few women who made a name for themselves in the Enlightenment, from Madame d'Épinay to Mary Wollstonecraft, came from rich families and were educated at home. It is hardly surprising, then, that the arguments and thought patterns of the enlighteners were still infused with such theological structures.

Nonetheless, even the demands and conclusions of the moderate enlighteners were daring for their time, since their philosophical, historical and scientific arguments always called political power into question too. After all, both the aristocracy and the church drew their legitimacy from a divine mandate and divine grace (rich Calvinists had also learnt to view their wealth as proof of God's grace, which simultaneously allowed them not to feel responsible for the poor). Thus, any argument that questioned the divine order and removed the authority of knowledge and morality from the crown and the church was an act of revolution in itself.

While the established powers eyed the energy of Enlightenment thought with great suspicion, the arguments of the enlighteners spoke all the more strongly to the middle class: the protagonists of the Industrial Revolution, of scholarly debate, of science, of trade, of political and social circles. The middle class could not base its claim to political power and a say in decisions on descent or the Bible, but it could invoke ancient philosophical arguments that had regained power in the course of this social revolution – namely, the right to freedom and human equality.

These principles are so strongly rooted in the thinking of modern people that they are taken for granted. In the seventeenth century, they were a moral scandal. Any decent person knew that there was a natural hierarchy between Christians and heathens, aristocrats and peasants, men and women, 'civilized' and 'primitive' peoples. The assertion of human equality was an assault on the natural order of society. The divine order was represented as a strict hierarchy.

Even the young Immanuel Kant was so sure of his theological position that, in the preamble to his first extended essay, *Universal Natural History*, he wrote that nothing could change his opinion, even if someone were to prove to him that the universe contained nothing but 'matter left to its general laws of motion' and the 'blind mechanism of the powers of nature' and relied on these to reach its form: 'If I had found this objection well-founded, the conviction I have regarding the infallibility of divine truths is so powerful in me that I would consider everything that contradicts them to be sufficiently disproved and reject it.'[47]

No proof would have been sufficient to change Kant's mind when it came to divine providence. Kant himself considerably changed his position over the course of his intellectual career, but the shadow of theology never disappeared entirely, even from his boldest ideas. His epistemology stated that all we can know about the physical world is what it is prepared to reveal to us and what we have the senses to perceive – namely, phenomena. Behind these, the Königsberg scholar prophesied, one would find the 'things in themselves'.

This is the ambivalence of the Enlightenment: on the one hand, Kant drove his contemporaries to despair because his philosophy declared that it was impossible by means of sensuous experience to perceive any of the world's 'essence' or a hoped-for spiritual truth – namely, God. On the other hand, like Descartes before him with his *res cogitans*, Kant created a space in which there was room for both the mystery and the creator, and which would never overlap with science.

Although Kant himself admitted that his model of the world was atomistic, like that of Lucretius, and thus left no room for God in its understanding of matter, he had created a final refuge for the deity in the 'things in themselves' and simultaneously created, with technical and analytical virtuosity, a mirror image of the Platonic idea that behind the world of phenomena there was another, and that this was the true world. The uncompromising Scot David Hume, Kant's great role model in these matters, had simply omitted the 'true' realm behind the phenomena and concluded from this that we can only ever deal with appearances, and have no justified, logical or empirical reason to suppose that there is anything else behind them; but this made the thinker from Königsberg shudder.

In a sense, Kant and other Enlightenment thinkers could not escape the trap that he himself had described as follows in his popular essay 'What Is Enlightenment?':

Thus it is difficult for each separate individual to work his way out of the immaturity which has become almost second nature to him. He has even grown fond of it and is really incapable for the time being of using his own understanding, because he was never allowed to make the attempt. Dogmas and formulas, those mechanical instruments for rational use (or rather misuse) of his natural endowments, are the ball and chain of his permanent immaturity. And if anyone did throw them off, he would still be uncertain about jumping over even the narrowest of trenches, for he would be unaccustomed to free movement of this kind.[48]

The 'immaturity which has become almost second nature', the shackles to which the spirit has grown so accustomed that it can jump farther with them than without them – Kant was probably writing from personal experience. He had freed himself from the strict faith of his parents, an act that was personally and intellectually revolutionary, but, more importantly, a moral act of personal self-empowerment. And yet his attachment to religion was still contained in the famous definition that opens this fine text: Enlightenment is man's emergence from his self-incurred immaturity.

Kant himself immediately asks why this immaturity is self-incurred, but his answer – namely, that humans simply do not use their understanding and prefer to think in conventional terms – is not entirely convincing. It is his religious upbringing that is significant here, for his family followed Pietism, a Protestant movement. The Pietists also practised adult baptism, since only adults could fulfil the divine mission of emerging from their ignorance and recognizing God. Whoever failed to do so incurred guilt, and could not be redeemed.

If one casts a vigilant eye on theology (which was far from uniform between confessions, and influenced its children in different ways), enlightened reason also appears in a new light – or, rather, in an old one. Although reason was contrasted with faith and superstition in particular, it was as similar to a central theological concept as two peas in a pod. For Enlightenment philosophers such as Kant, reason was the noble, immaterial part of humans that had to be emancipated and followed in order to control and overcome irrational physical desire and needs, in both individuals and entire societies.

This mechanism obviously resembles the Christian soul, which can only be liberated if lust and instinct, in a classically Neoplatonic gesture, are suppressed or sublimated and subordinated to its salvation – a domination of one's own unacceptable und thus guilt-laden inner nature. But if the good old soul sticks its nose out again from behind the elaborately painted scenery of enlightened, purely rational thought, then one must ask how much of the body of theological thought and supposedly long-forgotten discussions are still lurking behind this facade.

Borne by the optimism of the new sciences, the Enlightenment philosophers rode the wave of progress and the infinite improvability of the world. For a reading audience – even for Professor Kant – the vision of a history proceeding towards perfection was not only pleasant and flattering, but also matched the religious and philosophical ideas and narratives with which they had grown up. Progress as salvation history: it was easier to give an old idea an unfamiliar name than to think through an unfamiliar idea to the end. 'All men of the Enlightenment were cuckoos in the Christian nest', Peter Gay observed.[49] But perhaps they were not cuckoos' eggs that had been foisted on the church at all; perhaps they saw themselves as much more strange, much more different, than they seem in retrospect.

One finds more and more theology tumbling out of the scenery of moderate Enlightenment. One of the most intensely debated concepts in Christian theology is freedom of will, for, on the one hand, there cannot be sin and thus forgiveness if humans are not free to sin; on the other hand, it is difficult to reconcile this freedom with the omnipotence of God. One finds this discussion in the Enlightenment too: on the one side, the materialists who understood the entire universe as a clockwork and considered freedom of will pointless in a mechanistic world, and on the other side, the proponents of ethical freedom, who clung to it because there was no other way for humans to emerge from their self-incurred immaturity.

The duality of body and soul, the things-in-themselves, reason, progress, freedom of will – time and again, one finds motifs among Enlightenment thinkers that refer to a long Christian tradition and which, as with Kant, often seem to be taken up uncritically into their own thought. The last important motif in this series is the position of humans outside nature as its rulers. Bacon and Descartes had laid the argumentative ground for this. In the eighteenth century, the scientific domination of nature became not merely the road to a new Jerusalem, but also the Good News for entire societies.

A Work of Nature

The idea of humans as conquerors of nature on a divine mission, of virtuous humans gaining access to spaces and resources – by force if necessary – to fulfil the will of God or providence, was also crucial to the founding of the United States of America and the accompanying land appropriation. John Winthrop (1587/8–1649), the first governor of the Massachusetts Bay Colony, had already created the necessary theological framework. The land that 'lies common and hath never been replenished or subdued is free to any that will possess and improve it'.[50] The pastor John Cotton also argued that land only truly becomes property through cultivation. Without agriculture, the prairies and mountains of the continent were nothing but 'vacant soil'.

Over a century later, Thomas Jefferson reflected on the fate of indigenous peoples. The 'melancholy sequel' to their history, he noted, was the collapse of their population to a third of what it had been when the Europeans arrived. 'Spirituous liquors, the small-pox, war, and an abridgment of territory, to a people who lived principally on the spontaneous productions of nature, had committed terrible havoc among them.'[51]

Jefferson added that the land purchases had come about entirely legally, not through conquest. Nonetheless, a tribe now consisted of 'three or four men only, and they have more negro than Indian blood in them'.[52] Their lands amounted to no more than 20 hectares. Jefferson knew who was responsible for this sad state of affairs: the indigenous inhabitants had become slaves voluntarily.

Such official positions turned the displacement and extermination of other cultures into a moral mission, creating a happy combination of ethical good with economic and political utility. Naturally, there were also contemporaries who had always rejected and criticized the Western, church-sanctioned project of appropriating land and conquering other, 'less civilized' societies that had remained in the 'natural state', and who described their own presence with remarkable moral clarity.

The satirical novelist Jonathan Swift painted a picture of colonial conquests that he may have considered exaggerated, but which actually describes a reality expressed by many historical documents. A pirate ship goes off course and discovers an island by chance:

> they go on Shore to rob and plunder; they see an harmless People, are entertained with Kindness, they give the Country a new Name, they take formal Possession of it for the King, they set up a rotten Plank or a Stone for a Memorial, they murder two or three Dozen of the Natives, bring away a Couple more by Force for a Sample, return home, and get their Pardon. Here commences a new Dominion acquired with a Title by *Divine Right*. Ships are sent with the first Opportunity; the Natives driven out or destroyed, their Princes tortured to discover their Gold; a free Licence given to all Acts of Inhumanity and Lust; the Earth reeking with the Blood of its Inhabitants: And this execrable Crew of Butchers employed in so pious an Expedition, is a *modern Colony* sent to convert and civilize an idolatrous and barbarous people.[53]

Swift's biting irony is one example of many showing how the mentality of domination and the justifications for brutal oppression already made some enemies in the eighteenth century (and earlier). In France, Denis Diderot argued and railed passionately against colonialism, but had to do so anonymously in the book of a less renowned colleague to evade censorship, since voices opposed to the state interest in the ever contested colonies were not tolerated.

Other positions gained substantial publicity, but were thoroughly dishonest. Voltaire described movingly from his Swiss exile how the blood of slaves clung to each sack of sugar, though this did not prevent him from investing in those very sugar plantations. Nonetheless, public opinion, which expressed itself in ever more cheap publications and newspapers, could barely be controlled efficiently any more. The cafes and taverns of Paris, London and Naples were full of more or less dubious characters – private tutors, journalists, polemicists, materialists and hungry *abbés* whose clear-sighted cynicism about power was often born of their own painful experiences with the police and censorship. These figures were the fermenting mass for debate and for the opposition, which became part of European societies and from whose lightless corners some of the most important *Lumières* had emerged. This anarchic background is also reflected in an anarchy of opinions that were only directed into more orderly channels in the public debate through the formation of camps and the shaping of careers.

When engaging with these philosophical debates, one should always bear in mind that in societies monitored through censorship and inquisition, voicing opinions in public constituted a personal risk that could destroy an existence in the worst case. Inevitably, the multiplicity of positions was drastically reduced when they were conveyed to the public with names, powdered faces and royal printing privileges.

Nonetheless, there was an amazing range of possibilities for rethinking the relationship between humans and their natural surroundings, between 'culture' and 'nature'. Such positions often had no function in debates, which were heavily focused on justifying existing structures and religious dogmas, but they formed their own clandestine landscape, whose most important peaks were found not in officially published works of philosophy, but rather in novels, stories and plays, in personal letters and behind the closed doors of the salon.

The Enlightenment was never a philosophical school with its own catechism, even though it was often described as such in the nineteenth and twentieth centuries. It was never uniform in its thinking or in the themes of its passionate debates, which often varied from one country and one language to the next. It was always a programme whose content no one could really agree on, an ambition with uncertain aims. Most of all, however, it was full of innumerable contradictions, often even in the work of a single thinker. It was a lively debate that was in a state of growth and could not always control its energies.

The moderate Enlightenment, which clung to a form of creator or supreme being, was a difficult target for the authorities. It emerged from the bourgeoisie (except for aristocrats such as Montesquieu or the Comte de Buffon) and thought in figures and images that not only were acceptable for liberal citizens, but also provided arguments for their own ambitions and attitudes. Printers all over Europe and America supplied a flourishing market with physico-theological treatises, little booklets for ladies, edifying poems and multi-volume scientific disquisitions that went to great lengths to prove over and over again that there was no contradiction between science and religion – that religion already contained all science and had prophesied that religion is a product of reason itself, and so on. This Enlightenment sought to make peace with religious dogma and constructed the most adventurous bridges to enable the kind of coexistence already envisaged by Descartes. Science concerns itself with the *res extensa*, while religion (and philosophy, if it is well behaved) keeps the ensouled *res cogitans* to itself; any conflict is thus ruled out.

Naturally, these debates about the relationship between humans and nature and the essence of nature itself were never entirely well ordered or frictionless. Naturally, it was impossible to separate science from religion; the flood of archaeological finds and scientific studies had already made that impossible. By 1770, countless fossils had been discovered, countless well-preserved fossil seabeds, that were now located in mountain ranges far inland and indicated that land masses had once been covered by oceans and inhabited by unknown, extinct animals and plants, and the rock layers above these fossil seabeds already ruled out the notion that the Earth was 6,000 years old and simultaneously suggested that the creation was not finished after 6 days, as described in the Bible, but was rather an open-ended process still taking place.

In spite of all attempts to prove the contrary, these scientific results posed a direct threat to religious truth, and other results from research fields such as comparative anatomy, zoology, the description of hitherto unknown peoples, animals, plants and whole continents, called the official truth from the faculties and pulpits into question.

While the authors whom Jonathan Israel somewhat contemptuously, though not inaccurately, classes among the 'moderate mainstream' had to perform varyingly extreme contortions to reconcile their biblical fidelity with the newest scientific insights (or vice versa), there were always others who drew far more radical conclusions. The rediscovery of the Roman materialist philosopher Lucretius in the Renaissance contributed to this as much as the unideological clear-sightedness of a Michel de Montaigne, the ethics of a Baruch de Spinoza and the *Dictionnaire* of the French free spirit Pierre Bayle, who first made several 'heretical' thinkers and their arguments accessible to a small public at the end of the seventeenth century (allegedly in order to condemn them).

There were points of contact with other ways to think about the relationship between humans and nature, and in the eighteenth century a new generation embarked on a new intellectual outing to a *terra incognita* that could not be reached by any ship or colonized by any army: the unknown landscape of a world without matter and movement, without creator, hierarchy or goal. These ideas were explosive enough to make entire worlds collapse – and they still are.

Around the age of nineteen, David Hume suffered a nervous breakdown when he recognized the full implications of his own ideas and thought that he would henceforth have to live 'like a leper'. The notorious author of the materialist treatise *Man a Machine*, Julien Offray de La Mettrie, had studied theology for years in order to become a priest, a goal that had also brought Denis Diderot, a boy from the provinces, to Paris. Jean-Jacques Rousseau converted from Calvinism to Catholicism and back again; several authors of subversive works, such as Guillaume Raynal or Ferdinando Galiani, had taken minor holy orders and called themselves 'Abbé'; and the text that dismantled the church and religion as such in the bitterest atheist terms was penned by the priest Jean Meslier. His *Testament of Jean Meslier* could not be printed, but the manuscript circulated among intellectuals like samizdat literature in the Soviet Union.

Only a few people summoned the courage to stand by convictions that were unacceptable to society at large, and to live with the sometimes drastic consequences, and few were able to make their arguments public; for a period of many centuries, opinions deviating from the consensus have been preserved mainly in the court records from heretics' trials. Such statements, often extracted through torture, are among the few sources testifying to different views in the history of Christian Europe. The worldviews articulated by the accused are often isolated and not philosophically elaborated; many of them were simple people or had access to clandestine literature, but European thought was also more diverse than the libraries full of theological texts and censor-approved volumes would lead us to believe.

The radical thought that emerged in the second half of the eighteenth century, especially in Paris, developed a conception of the world which was considered so dangerous and subversive that the mere possession of a book could be a death sentence; recall the poor Chevalier de la Barre, who was gruesomely executed in 1766 at the age of twenty-one after the police found Voltaire's *Philosophical Dictionary* in his room.

Undoubtedly the most important of these authors, a French author of German heritage called Paul Henry Thiry d'Holbach (1723–89), was only able to publish his great philosophical bulldozer, written with a certain stubbornness, in the first place because he had his anonymized manuscripts smuggled from Paris to Amsterdam, where they were published under a pseudonym in a substantially more liberal climate, then imported back to France hidden in hay bales and herring barrels.

It is not difficult to see why d'Holbach's thinking posed such a threat. In his central work *The System of Nature* there is no place for God to hide, not even in the most diluted and metaphorical form. It opens with the following words: 'The source of man's unhappiness is his ignorance of Nature. The pertinacity with which he clings to blind opinions [...] appears to doom him to continual error.'[54] From childhood, d'Holbach argued, humans are forced to wear a blindfold; he made it his task to pull it off.

> [Man] is the work of Nature. He exists in Nature. He is submitted to the laws of Nature. He cannot deliver himself from them: cannot step beyond them even in thought. It is in vain his mind would spring forward beyond the visible world: direful and imperious necessity ever compels his return – being formed by Nature, he is circumscribed by her laws; there exists nothing beyond the great whole of which he forms a part, of which he experiences the influence. The beings his fancy pictures as above nature, or distinguished from her, are always chimeras formed after that which he has already seen, but of which it is utterly impossible he should ever form any finished idea, either as to the place they occupy, or their manner of acting – for him there is not, there can be nothing out of that Nature which includes all beings.[55]

D'Holbach's somewhat hammering tone becomes a challenge over the course of 700 pages, but the immensely high circulation that his works already achieved in the eighteenth century despite the strict censorship, or perhaps because of it, made him one of the most influential thinkers of his time. He had trained as a lawyer, but refrained from entering the profession after inheriting a fortune from an uncle. Instead, he wrote forbidden books, discreetly helped artists and authors in need, paid for translations of works he considered interesting and hosted one of the most important salons of the time.

It was frequented by some of the most brilliant minds of the period, from the perennial literary star Denis Diderot to the Scot David Hume and the Neapolitan Ferdinando Galiani. Jean-Jacques Rousseau was a regular guest before falling out with his old friends, as were the zoologist Georges-Louis Leclerc de Buffon, the intellectual *salonnière* Madame de Geoffrin and a strong delegation from across the English Channel including the moral philosopher Adam Smith, the actor, rediscoverer of Shakespeare and wine merchant David Garrick and the writer Laurence Sterne.

The host assembled an astonishing galaxy of minds in his elegant but simple city house not far from the Louvre, providing a space for debates without the police informants lurking outside: a society in which everything could be discussed – and was. Those present included scholars and scientists, some of them authors of the encyclopaedia edited by Diderot, whose goal was to collect the entirety of useful knowledge of their time. This undertaking was doomed to failure, as Diderot knew very well, but it not only had an enormous symbolic effect – it also paid his rent.

D'Holbach's wealth allowed him to devote his life to his intellectual passion, to write works and circulate ideas in a way that would scarcely have been possible for less privileged contemporaries. His philosophically uncompromising nature was legendary. His personal secret was that, as Max Weber later said of himself, he was religiously unmusical. He saw a physical world consisting of nothing but matter and movement. All phenomena could be explained in terms of these, even if the current state of scientific knowledge was not yet sufficient to do so. Unlike Diderot, who still mourned the lost faith of his childhood in old age and was thus able to write with empathy about the conflicts of a doubter, d'Holbach does not seem to have been assailed by any such doubts.

Materialist thought was attacked by its many opponents as human megalomania and a world without morals, but its aim was precisely the opposite. In a philosophy that acknowledged nothing outside or above nature, the idea of dominating nature was also untenable. As a 'work of nature', humans had no choice but to follow its laws and use science to understand them better, and thus benefit from them.

The moral dimension of this way of thinking developed entirely of its own accord: desire and empathy are part of human nature, and, just as eros makes humans seek and need closeness to other humans, empathy causes them to feel the pain of others. To achieve happiness, one must be surrounded by happy people, which requires the reduction of suffering. Thus, the Christian concept of virtue was redefined:

> Thus *virtue* is every thing that is truly beneficial, every thing that is constantly useful to the individuals of the human race, living together in society; *vice* every thing that is really prejudicial, everything that is permanently injurious to them. [...] The man who injures others, is wicked; the man who injures himself, is an imprudent being, who neither has a knowledge of reason, of his own peculiar interests, nor of truth.[56]

However, this natural social order is rendered impossible by a perverse doctrine that forces humans to live contrary to their own nature and thus cripples them morally and intellectually. Baron d'Holbach's argumentation was in step with his time and its travel reports:

> We are informed, that the savages, to flatten the heads of their children, squeeze them between two boards, by that means preventing them from taking the shape designed for them by Nature. It is pretty nearly the same thing with the institutions of man; they commonly conspire to counteract Nature, to constrain and divert, to extinguish the impulse Nature has given him, to substitute others which are the source of all his misfortunes.[57]

These lies infantilize humans, keeping them under control with illusions. The boards around our own heads may be metaphorical, but this makes it all the more difficult to recognize and remove them.

But there were also more subtle ways to think within and through nature than d'Holbach with his slightly formulaic certainties. His friend and interlocutor Diderot likewise saw humans as natural beings in a natural world, but they were full of contradictions and conflicts even without the deformations caused by society. Head, heart and loins often pull in different directions, and it is a constant challenge to find a balance between them.

Diderot often articulated his best philosophical ideas in letters or literary works. His novella *D'Alembert's Dream* is a materialist manifesto (which is why it was only published in 1830, forty-three years after his death), in which one of the protagonists develops a fascinating theory:

a human being is 'a machine that builds itself little by little through a multitude of successive stages; a machine whose regularity or irregularity is determined by a packet of fine, loose, flexible threads and by a kind of embroidery frame where the smallest thread cannot be crushed, broken, displaced or removed without the most disastrous consequences for the whole organism.'[58] The man passes on a fine thread (*brin*) to the woman during coitus, and it combines with a fine thread in the woman. This process results in mistakes if the threads are 'crushed, broken, displaced', he replies, when asked why children resemble their parents but are not identical to them.

This speculative genetics, which Diderot committed to paper in 1769, was augmented by a further, no less daring idea: humans are nothing but thinking, feeling matter, consisting of smaller units that are likewise capable of thinking and feeling – just as a swarm of wild bees on a branch forms a kind of body that moves and changes, but comprises millions of mutually coordinated individuals. All matter is connected, like an immense spider web; nothing exists only for itself. Life and death are nothing but states of an ever changing matter. Marble can become flesh if marble dust nourishes plants that are subsequently eaten. Nothing lasts – it is all part of the endless chain of being. But Diderot would not have been Diderot if he had made peace with this implacable idea. He wrote to Sophie Volland, his lover of many years:

> Oh my Sophie, so I still might hope to touch you, to feel you, to love you, to approach you, to unite and mingle with you when we are gone! If only there were a law of affinity between the elements of which we are composed if we were destined to become one single being, if in the course of centuries I were to become one with you, if the scattered molecules of your lover could live and move and search out your molecules dispersed through nature! Do not take this fancy away from me; it is dear to me, for it would give me the certainty of living eternally in you and with you.[59]

The radical thinkers who gathered in d'Holbach's salon referred to an intellectual tradition that they traced back to classical antiquity, and which had a different approach to the relationship between humans and nature. This tradition had always been peripheral to European thought, for it had a severe disadvantage: thinking from the perspective of nature, without reference to any fixed point of transcendence, holy text or revelation, along with the multitude of accompanying exegetes, was completely unsuited to legitimizing the power of rulers.

This was also expressed in the political ideas of those thinkers. Diderot began life as a constitutional monarchist, but by the end of his life he had moved towards an anarchism that saw all forms of power as problematic. D'Holbach and allies such as Claude Adrien Helvétius and Nicolas de Condorcet searched for forms of a republic of virtue, which was scarcely available as a realistic political vision, but certainly meant thinking thoroughly about social and political aims. Their interlocutors from previous generations – Socrates and Lucretius, Seneca, Niccolò Machiavelli, the Thomas Hobbes of *Leviathan*, Hugo Grotius – had different ideas about justice, the legitimacy of power, virtue and human nature, but what they had in common was that they sought the answers to those questions within nature and society. Precisely someone like Machiavelli was unconcerned about legitimacy bestowed from beyond, and many of his contemporaries clearly felt the same. It was therefore more useful to be feared than loved. Power is a machine, and machines need control; otherwise, they must be taken apart and reassembled.

One name is missing from this little homage to d'Holbach's salon, although he was a frequent and important guest: Jean-Jacques Rousseau, whose intellectual kinship and deep friendship with Diderot ended when he broke both philosophically and personally with all his old friends. Growing up in the strictly Calvinist Geneva as the son of a tyrannical father, Rousseau fled from his home town while still a teenager, which was an immense act of personal liberation and the prelude to an inconstant and ultimately lonely life.

Maybe it was this biographical dimension that turned Rousseau into such an incisive critic of political power. He was still a young man when he wrote his *Discourse on the Origin of Inequality*; the paranoid states that later befell him had not yet manifested themselves, and his spirit was torn between his personal and intellectual rebellion, the new scientific discoveries of his time and a deep longing for spiritual security and the religious sentiment of his childhood. His ambition was to understand the foundations of the inequality he observed everywhere.

Rousseau began his search in a remarkably empirical way. If one takes away culture and 'artificial faculties', it becomes clear: 'when I consider him [man], in a word, as he must have left the hands of nature, I see an animal less strong than some, less agile than others, but all in all, the most advantageously organized of all'.[60]

The human being: an 'animal' that was anatomically lucky? Rousseau imagines 'him' drinking from a stream and then falling asleep under an oak – a pastoral idyll, a kind of Enkidu of the Enlightenment imagination concealing a materialistic nature: 'Every animal has ideas, since it has senses; up to a certain point it even combines its ideas, and in this regard man differs from an animal only in degree.'[61]

This 'natural state' of humans was characterized by simple needs; they lived in loose polyamorous groups and were neither good nor bad, but simply followed the laws of nature. All that came to an end, however, when it finally occurred to a human to take God's assignment to Adam and Eve seriously, inventing all the vices of the bourgeoisie as a result: 'The first person who, having enclosed a plot of land, took it into his head to say *this is mine* and found people simple enough to believe him, was the true founder of civil society.'[62]

Karl Marx would also adopt the vision of a primordial, primitive communism postulated by Rousseau (and Adam Smith, incidentally) as the beginning of human history: the idyllic antithesis of the Industrial Revolution's sinister dynamic. In fact, there is no anthropological evidence that tribal societies of hunter-gatherers ever lived in that way. For example, even in tribal societies (as among other primates and mammals), hunters had privileged access to quarry they had killed and its distribution among particular places and territories. Essential tools like weapons are also often personal property. Such structures have been documented independently among peoples in South America, Alaska, southern Africa and the Philippines.

In the eighteenth century, authors were not yet troubled in their view of history by such inconvenient facts, and could therefore devote themselves entirely to a speculative history that confirmed their ideas about the present. Thus, Rousseau describes the consequences of this far-sighted decision to appropriate land – and the lack of resistance – in the following terms:

What crimes, wars, murders, what miseries and horrors would the human race have been spared, had someone pulled up the stakes or filled in the ditch

167

and cried out to his fellow men: 'Do not listen to this impostor. You are lost if you forget that the fruits of the earth belong to all and the earth to no one!'[63]

It was too late; property had been invented, and with it greed, oppression and decadence. The only solution for this terrible spiral of degeneration and brutalization is a return to the laws of nature. Here Rousseau was too good a thinker to believe his own arguments. Is a return to a natural state ever possible? And who defines what that natural state was? And who supposedly has the power to implement these profound insights politically?

Rousseau's early writings make it clear eighteenth-century thinkers already saw the divine mission given to Adam and Eve as a flimsy pretext for political ambitions. The subjugators always need simple-minded people who allow themselves to be subjugated – an observation made by the prematurely deceased Étienne de La Boétie (Montaigne's intensely mourned lover) in his polemic *Discourse on Voluntary Servitude*.

Rousseau's voice stood out from all the paeans to reason, culture, science and progress. To him, what others called culture was nothing but the perversion of natural instincts, and the metropole as the incarnation of the new society was a place of vices, decadence and the destruction of true feelings. This polemic was the starting point for an intense, often politically ambivalent tradition of Romantic, counter-Enlightenment thought. Man, the natural being perverted by culture, which can only find inner peace and its true destiny in the silence of the forest, on the steaming soil or in the expanses of untouched landscapes, preoccupied writers and poets from Novalis to Rilke, from Emerson in America to Coleridge in England.

Thanks to its almost unlimited flexibility, it could be used to criticize the enlightened bourgeoisie and later as a motivation for the First World War – the rebirth of the true man after being softened by urban life – and as a basis for fascist ideologies, from the Nazis to Vladimir Putin's fantasy of an original, pure, unspoilt medieval Rus. But such phenomena as the life reform movement around 1900 and the hippies, who opposed the triumphant post-war consumer societies and their proxy wars with a different vision of life, or the environmental movement in the late twentieth century, would also have been almost inconceivable without Rousseau.

In the Romantic worldview, humans were and always would be part of nature, alienated from their origins by their own delusions – and were afflicted by this. Healing this affliction involved a return to the purity of the beginning and the suppression (potentially a violent one, even for Rousseau) of all false, foreign and mendacious tendencies and their agents. All too often, what lurked behind the dream of pure community was the reality of dictatorship.

Virtuous Terror

Perhaps the danger of a dictatorship is inherent in the Enlightenment's own internal dynamic, and there is a direct path leading from the rule of reason and the subjugation of nature and all things natural to the reign of terror of a Maximilien Robespierre, who sent his opponents to the guillotine in the name of reason and virtue.

One rarely found such immediate brutality among Enlightenment thinkers, who saw themselves above all as the intellectual vanguard of a new world. Even enlightened social utopias have unmistakable tendencies towards a subjugation of the natural, however, as shown most clearly by the panopticon invented by Jeremy Bentham (1748–1832).

Bentham, a wealthy private scholar, would probably be diagnosed with autism today. Throughout his long life he found it difficult to understand the emotions of others, lived by a strictly regulated daily routine and was considered intellectually extreme, but with a very specific talent. He began learning Latin at the tender age of three, and was sent to Oxford by his father when he was twelve. Bentham never stopped marvelling at the irrationality of human behaviour. His own moral philosophy was simple, logical and coherent: 'It is the greatest happiness of the greatest number of people that is the measure of right and wrong.'[64]

Whatever makes most people in a society happy is right by definition. His critics countered that, based on this criterion, there would have been nothing objectionable about Christians being thrown to the lions in ancient Rome, since the majority of people found it entertaining; but Bentham was undeterred. He argued that there are only two principles in nature:

> Nature has placed mankind under the governance of two sovereign masters, *pain* and *pleasure*. It is for them alone to point out what we ought to do, as well as to determine what we shall do. On the one hand the standard of right and wrong, on the other the chain of causes and effects, are fastened to their throne. They govern us in all we do, in all we say, in all we think.[65]

To develop a good society, one must adapt the principle of pleasure and pain to useful social principles and thus support rational, virtuous and useful behaviour. This had important consequences; Bentham referred to the idea of 'natural law' – which had been so important in philosophical discussions and assumed that certain rights exist by nature, or through the creation – as 'nonsense on stilts'. Rights do not exist, he argued, but are created when people bestow them on one another. This means that they have no rights over other creatures, especially since humans are more similar to them than they might like:

> The day may come when the rest of the animal creation may acquire those rights which never could have been withholden from them but by the hand of tyranny. [...] But a fullgrown horse or dog is beyond comparison a more rational, as well as a more conversable animal, than an infant of a day, or a week, or even a month, old. But suppose they were otherwise, what would it avail? The question is not, Can they *reason*? nor Can they *talk*? but, *Can they suffer?*[66]

In a very English manner, Bentham's sincere empathy with other animals excluded the two-legged type; while he recognized animal suffering as a moral problem, his projects for human societies were resolutely defined by their utility for the greatest number of people, not by empathy with individuals.

The most famous project associated with Bentham's name was never realized in his lifetime and mostly floats about as a metaphor. A trip to Russia gave Bentham the idea for his most ambitious project, which he would doggedly pursue for years. The idea came to him in 1787 in Krychaw during a visit to his brother Samuel, who was working there for Prince Potemkin, of all people – the famous favourite of Catherine the Great and inventor of the Potemkin villages, which consisted only of facades that the prince could sledge past with his tsarina to present an illusion of progress. They were probably never built, but *se non è vero, è ben trovato*.[67]

Perhaps it was the reformist ideas that were in fashion in Russia at the time, and to which Potemkin devoted such energy, as well as the influence of the French Enlightenment thinkers at the tsar's court, or maybe it was the boredom Bentham felt while far from home that gave him his new idea; at any rate, he soon felt that he had stumbled on an ingenious inspiration, a simple and secure method of not only addressing but completely eliminating all possible ills in society: 'Morals reformed – health preserved – industry invigorated – instruction diffused – public burthens lightened – Economy seated as it were upon a rock – the Gordian knot of the Poor-Laws not cut but untied – all by a simple idea in Architecture!'[68] He called this idea the panopticon.

The panopticon seemed compatible with every purpose:

> whether it be, that of *punishing the incorrigible, guarding the insane, reforming the vitious, confining the suspected, employing the idle, maintaining the helpless, curing the sick, instructing the willing* in any branch of industry, or *training the rising race* in the path of *education*: in a word whether it be applied to the purposes of *perpetual prisons* in the room of death, or *prisons for confinement* before trial, *penitentiary-houses*, or *houses of correction*, or *work-houses*, or *manufactories*, or *madhouses*, or *hospitals*, or *schools*.[69]

This ingenious idea was strikingly simple: a large circular building with cells along the outer wall, and in the centre a watchtower separated from the cells by a ring-shaped empty zone. An important aspect was that on the inward-facing side, the cells only had bars; thus, the warden in the tower could always look inside them, but remained unseen by inmates. As a result, no inmate could ever know whether they were being observed or not.

Bentham had calculated everything precisely, from the diameter of the building (100 feet) and the thickness of the walls to the size of the cells and their windows, from the illumination and communication to the food, daily schedule, punishments, activities and, naturally, the economic viability of the overall project, since the panopticon was meant to be self-sustaining and even profitable, so that honest citizens would not have to pay anything for it. To make his ideas more tangible, he commissioned an architect to draw the building he had conceived so meticulously.

The panopticon was a machine of social transformation. In the eighteenth century, prisoners often languished under frightful conditions, in dark, damp cells, without adequate sustenance; they were exposed to heat and cold, to illnesses and the brutality of their wardens. For them, the panopticon would certainly have been an enormous step forwards: a rational regime without tyranny in which they would have been able to work for a living.

Anyone who was not willing to work would be subject to strong sanctions. Here Bentham's moral enthusiasm had an important precursor, even if the philosopher did not realize it: another architectural legend that may never have existed, but which is as much an expression of the culture that gave birth to it as Potemkin's unbuilt villages. In the seventeenth century, the Rasphuis in Amsterdam was a special prison for the work-averse and 'antisocial elements', whose purpose was to integrate prisoners back into society through work. Whoever refused was persuaded with blows from a stick; according to local legend, there was a special place for those who were especially stubborn: the drowning cell. The prisoner was placed alone in a windowless room that slowly filled with water. All he had to save himself from death was a pump, so he had to pump incessantly to avoid drowning.

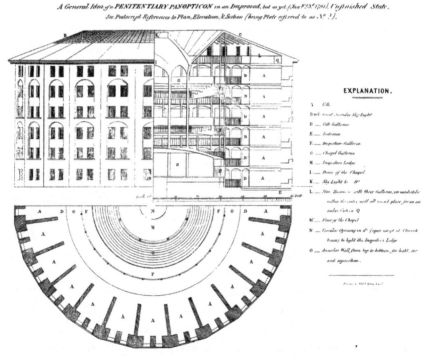

11 Jeremy Bentham's panopticon, drawing by Willey Reveley (1791)

Bentham too believed in the healing value of work, but he also had other ideas about what to do with the prisoners. One could use them to test not only working methods, but also medication and new forms of punishment. Children could be brought up in different ways in different panopticons to test pedagogical theories or psychological hypotheses, all for the good of society.

At the same time, the calculus of happiness on which the philosophy of the father of the panopticon was based also led him to positions that were very unusual in his time and still seem amazingly liberal to us today: he not only championed animal rights, but also demanded the legalization of homosexuality, gender equality and the abolition of the death penalty, which, he argued, disproportionately affected the poor and was more expensive to implement than the economically productive option of lifelong forced labour.

After returning to London, Bentham immersed himself in the realization of his project with intense enthusiasm. A building site was found and the government was not averse to it, but planning soon began to stall. Neighbours protested, political intrigues were an obstacle to implementation and the king, who felt insulted by the ever socially awkward Bentham, slowed down the project's progress until everything ground to a halt and the idea was ultimately abandoned. Bentham, who had put large sums of his own money into the planning, was beside himself with anger, and this anger soon turned into a profound, lifelong bitterness.

Thus, no panopticon was built in Bentham's lifetime, and he turned towards other projects. As a counterweight to the elitist and, in his view, useless universities of Oxford and Cambridge (he still had bad memories of the former), he founded University College in London, a place to which he felt deeply connected. He died in 1832 at the age of eighty-four.

In his will, Bentham specified not only that his body should be autopsied in the presence of his friends – this would allow the gentlemen to learn something from his death too – but also that it should be mummified, dressed in his everyday clothes and kept at the university in perpetuity. He can still be seen there today, though the head (which kept falling off and was stolen several times by students from other universities) has meanwhile been replaced by a wax model. For the university's 150th anniversary in 2013, his 'auto-icon', as he called it in his will, was brought to the ceremonial council meeting. The minutes state that the philosopher was 'present, but not voting'.

The panopticon has long become a metaphor for an all-pervasive, all-surveilling modernity. Michel Foucault wrote about this in his pioneering book *Discipline and Punish*, and today one could fill a small bookcase with works about Bentham's invention. The horror triggered by his idea had deeply troubled him. For example, he had drafted a law in 1794, the Panopticon Bill, which stipulated that all prisoners should have their name, birthplace and date of birth tattooed on their left arm. He responded with bafflement to the subsequent criticism and outrage with the remark that he would happily have led by example and had such a tattoo made himself. The depths and contradictions of human emotions remained a hidden mystery for the lifelong bachelor, whose only constant companion was a cat.

It was not until the twentieth century that Bentham's idea was put into practice in several places, usually modified to some extent. The Presidio Modelo on the Cuban island of Isla de la Juventud is one of the most faithful realizations of Bentham's plan; another is the F-House no. 2 at Stateville Correctional Centre in Crest Hill, Illinois – a huge antiseptic hall with a floor of polished concrete, steel bars, spotlights and the aesthetic of a cell on death row.[70]

*

What is commonly referred to today as 'the Enlightenment' was a complex, highly geographically variable and often contradictory form of intellectual reappropriation of a world that could no longer be adequately explained using religious ideas.

This was less a voluntary reappropriation than the result of a clear necessity, for the world of the seventeenth century was immensely larger, more complex and more in need of explanation than it had been 200 years previously. This world offered far more possibilities for action, especially for the urban bourgeoisie, the real agent of social and political change, whose emphasis on equality and universal (albeit not too universal) human rights was not least a way of philosophically buttressing its own interests.

Because this process was influenced by strong social interests from the start, it always triggered power struggles, extending to the most seemingly abstract argumentative ramification. The social interests and background of the participants varied considerably, however. Some authors were revolutionaries from the start, while others strove for reforms, and still others sought to save the theological and political status quo by making philosophical concessions to a rational, abstract creator in order to reject compromises in practical matters all the more resolutely. Many tried unsuccessfully to secure a place for God in the system of reason, while others accepted juxtapositions of mutually incompatible ideas without commentary. They all changed their minds in the course of debate depending on how much independence they could respectively afford, but almost no one could write what they really thought.

So was the Enlightenment really a triumph of reason or a cynical power game? A call for revolution and universal rights, or a deeply conservative project to cement old power structures and theological ideas with a new vocabulary? Did it open up new horizons or enable new exploitation?

All of the above, and much more. Its legacy remains ambivalent – not only for the reasons that critics such as Michel Foucault or Max Horkheimer and Theodor W. Adorno would present in the twentieth century, but because the moderate Enlightenment thinkers, from Descartes to Voltaire and Kant, were able to break away entirely neither from the God of their childhood nor from the associated theological ideas, or were unwilling to do so out of political motives.

One can find many methods of theological thought that were taught especially at Jesuit-run schools in Enlightenment polemics – such as the citation of ancient sources, except that a Roman author was chosen instead of the Gospel, or argumentation based on a natural order that could be uncannily similar to a divine one. But central elements of such a worldview were thrown largely unquestioned into these discussions and adopted in widely sold and read texts from this time. Many of the pillars of enlightened thought have direct parallels in the theology with which its exponents had grown up, and whose rhetoric was still central to their thinking.

Only the radical materialists dared to question the assumption that humans are fundamentally different from animals and ask whether humans should really occupy a special place in nature, and where exactly the soul is located in their bodies. Hardly any of them doubted that history moves towards a clear goal.

The arguments and authors from which the bourgeois historians of the nineteenth century then selected and edited 'the Enlightenment', which was packed into textbooks and taught in school, represented a particular branch in the immense intellectual delta of debates over two centuries. This branch of the wide river allowed reason to explore the world, subjugate it and explain it using a rationalist model, without calling into question the mystery of faith or the existence of a final guarantor for truth, morality and one's own self-image.

This moderate and regimented Enlightenment, purged of its opposing tendencies and discrepancies, was still explosive enough to profoundly undermine the dominance of the church and the nobility, and thus the structures of societies. The new bourgeois and, more or less, republican masters insisted that power was now distributed differently, based on democratic and rational criteria, and no longer required any divine grace to justify itself before the tribunal of humanity. However, it was precisely the modern Enlightenment figures that ensured a strong continuity of power, albeit now supported scientifically rather than theologically, affirming the place of humans (specifically the male European variety) at the top of creation and all natural hierarchies and assigning to them the special mission of bringing about their own liberation by spying on, outsmarting, overpowering and exploiting nature.

The decisive factor that made the difference between legitimate and illegitimate subjugation was, as Descartes had already shown, the presence of a soul – or, in the vocabulary of the Enlightenment, of fully valid reason. This distinguished culture from nature, subject from object, lawfulness from lawlessness. Deciding who or what could be described as part of nature became one of the most effective and most wicked instruments of power in history.

Carte Blanche

Land-appropriation thus is the archetype of a constitutive legal process exter-
nally (*vis-a-vis* other peoples) and internally (for the ordering of land and
property within a country).

Carl Schmitt[71]

Nature has made a race of workers, the Chinese race [...] a race of masters
and soldiers, the European race.

Ernest Renan[72]

Generations of children from Hamburg (this author included) have
poured through the famous gate of Hagenbeck Zoo. The complex was
inaugurated in 1907; the gate exudes the charm of stucco-decorated
bourgeois buildings and captains' villas in Altona, an emblem of a self-
confident era (figure 12). At the same time, the gate is a promise of an
entrance to another, exotic world symbolized by the wild animals, by the
polar bear, lion and elephant – and the other 'savages': the Eastern warrior
with a shield and spear and the Native American with a tomahawk and
gun, who seems to be letting out a bloodcurdling war cry.

Carl Hagenbeck's zoo was innovative in its day. The enclosures were
not surrounded by heavy iron bars, but rather crafted skilfully so that
the creatures locked up there could be viewed in a simulation of their
natural surroundings. The polar bears and penguins had an iceberg made
of plaster and wire mesh, the chamois climbed up and down artificial
hills, and the Indian elephants roamed through a Hindu temple with
altars. In an area of 19 hectares, a wide range of animal species was
displayed in the most faithfully rendered natural surroundings; it was
a voyage across continents that, like the Arab and the Native American
on the gate, had much in common with the novels of Karl May – an
almost magical experience in a time before television and long-distance
travel.

12 Hagenbeck Zoo in Hamburg, main entrance; postcard, 1919

And yet it had all begun very humbly with an ordinary fishmonger who, in 1848, had the superb idea of buying a few seals from the fishermen of St Pauli and displaying them at the market in wooden tubs for an admission fee of a few pfennigs. They were later joined by a polar bear and a hyena. This grew into a show business that the fishmonger's son, Carl, expanded into a zoo that financed entire expeditions to bring rare animals to Europe from all over the world. Hagenbeck himself worked at the fish shop as a boy and barely went to school. Nonetheless, he admired his father:

> He was a man of unshakable principles and noble views. I must say with great gratitude that he laid the foundation for everything that was achieved. His character combined a great seriousness about life with a good-natured manner. The outwardly strict way in which he brought up his children concealed a great warm-heartedness. The rod played no part in our upbringing; the model of our father, who was the embodiment of activity, punctuality and frugality, already taught us children to live in his spirit.[73]

Carl Hagenbeck's recollections are full of jovial humour and show with notable honesty how a modest show business could grow into

Europe's largest animal dealer and zoo, making him rich and famous. This profession demanded a passion for animals, he asserted, even if his relationships with his creatures were often based on a healthy business sense. When one of his elephants attacked a warden, he decided to have the animal 'executed' and sold this privilege to a hunter from England, who brought an entire arsenal of weapons to Hamburg but, at the decisive moment – the elephant had been staked to the ground outside, in front of a wooden wall – he was too nervous to pull the trigger. Because he had paid, however, he did not want anyone else to shoot the elephant bull. So Hagenbeck had an idea:

> The condemned giant was therefore driven back into the stable, and a noose was placed round his neck. The rope was wound round a pulley, attached to a cross-beam under the roof, and six of my men played the part of executioners. 'One, two, three' I called out, and at the third shout they all hauled on the end of the rope. The bull almost immediately lost the ground under his feet, his head fell sideways, and in less than a minute he was dead. We found afterwards that his neck had been broken. Thus ended one of the strangest tragicomedies which I have ever seen.[74]

In fact, Hagenbeck had to keep finding new sources of income, since trading in exotic animals was less profitable than he had hoped; the expeditions cost a fortune, and most animals died before even reaching Europe. But Hagenbeck still had an ace up his sleeve, a substantially more lucrative invention that he claimed to have made himself: ethnographic exhibitions to educate and entertain a large audience. Here the zoo director revelled in proud memories; his shows had travelled to London, Berlin, and in 1886 even to France: 'This ethnographic exhibition had been the sensation of Paris. It had not only provided substantial earnings, but had also brought entertainment, stimulation and education to an incalculable audience. On Sundays the show had drawn up to half a million visitors.'[75]

First of all, Hagenbeck had imported a few families from Lapland, along with their reindeer, to show them going about their daily lives. The short-statured Sami, with their reindeer-fur costumes and their weapons, tents and tools, were a sensation in Hamburg, and the businessman saw his chance to make a killing. He sent his human exhibits on tour all over the country and began to forge more ambitious plans. Next he would display a complete African village, with animals and people from Africa performing dances and rituals. He cast his net ever wider to offer the paying public increasingly sensational spectacles:

> My Singhalese troupe was veiled in the magical aura of the ancient wondrous India. We had not only captured its picturesque exterior, but also the shimmer of its mystique. The colourful, captivating sight of the camp, the majestic elephants, some of them draped with radiant gold saddlecloths, some of them dragging huge loads with harnesses; the Indian magicians and jugglers, the dancing devils with their grotesque masks, the slim, beautiful, doe-eyed bayaderes with their sensuous dances, and finally the great religious Perra-Harra procession – all this cast an entrancing spell that enthralled audiences everywhere.[76]

The exotic sensuality and an air of oriental magic aroused so much curiosity that the police had to intervene, to protect both people and animals. Hagenbeck came up with something even more cunning for his African presentations:

> At the beginning of the show, slave traders suddenly 'raided' this peaceful village. Accompanied by cries and gunfire, Arabs perched on dromedaries rode around the villagers, who had been feasting only moments before. The herds of goats dispersed, chickens fled amid loud clucking and, after a brief scuffle, the poor captives, bound very realistically with chains and wooden yokes, were taken away as live loot. Then European animal-catchers appeared and chased away the marauding Bedouins, and then there was a great peace festival at which, accompanied by native music, there was dancing as well as a performance of all the rites of a genuine Sudanese tribal celebration. This was followed by peacock hunts on racing dromedaries.[77]

Carl Hagenbeck writes warmly and with a certain sentimentality about the people from other continents who appeared in his ethnographic exhibitions, but leaves certain things unmentioned. Many of those who signed a contract with him were lured into making the journey with false promises of work and prosperity, and never returned home; they died of smallpox and tuberculosis. The Fuegians whom Hagenbeck presented in 1879 took the diseases back to their island; thirty years later, the complete population had been wiped out. The deportees were not vaccinated. Those who already died during the voyage were simply thrown overboard, while the survivors, having arrived in Europe, were transported in cattle trucks and declared at the borders in keeping with toll regulations. Those who survived all this had to endure the stares of daily visitors and perform the required programme until they were too sick to be exhibited. The death of a family member in the group was not accepted as a reason to be absent from a performance.

The ethnographic exhibitions were not Hagenbeck's own invention. The Jardin d'Acclimatisation in Paris had introduced the displays of 'exotic' peoples, and fairground stalls had exhibited 'freaks' and 'savages' for over a century, but the ethnographic exhibitions became a special phenomenon. They were soon taking place in hundreds of cities in Europe and the USA, especially at world exhibitions such as those in Vienna (1872), Chicago (1893) and Paris (1900), but also at colonial exhibitions. They gave visitors a taste of what awaited them beyond civilization, among the 'savages', whose seemingly primitive and violent existence could only be corrected by the blessings of colonial rule. It did not matter that the scenes and costumes in which they had to present themselves often had nothing to do with life in their countries of origin, that the dances and rituals had been invented for the stage. The staging was decided long before the journey began: 'primitive' but sensual Africans; proud and cruel Arabs; Native Americans with full feather headdresses doing the rain dance; mysterious Indians; primeval Fuegians, cannibals, wild and almost like animals – ideal for the zoo and the circus.

How was it possible for a man who prided himself on the unshakable principles he had inherited from his father to exhibit humans for profit and accept their deaths with the same smile as the 'execution' of an elephant?

The answer to this question can be found two centuries before the ethnographic exhibitions in the radiant Caribbean sun: in the journal of Thomas Phillips, captain of the African Company's ship *Hannibal*. Phillips was Welsh and no longer a young man. He wanted to settle down in his home country, and commanding a ship in the African Company was a great opportunity for him to finance a comfortable retirement. In 1693 he was tasked with sailing from Bristol to the western coast of Africa to pick up his cargo: human beings, whom he would buy at the local slave market in exchange for bulk goods such as knives, iron bars and tin pots, then sell to the owners of sugar plantations in the West Indies.

This 'triangular trade' was a tough business. The *Hannibal* was already attacked on the way to Africa, and only just managed to defend itself, despite its thirty-six cannons. It had to seek refuge at the next port so that the shot-up, half-dismasted and leaking ship could be repaired. Shortly afterwards, the ship and crew almost fell prey to a hurricane. Life on board was also difficult. Off Liberia, Phillips's evidently far younger brother (probably his half-brother) caught a fever that was spreading among the crew and died within a day. The captain wrote that it had left him 'filled with pain over the loss' and gave him military funeral honours with trumpets, drums, flags at half-mast and sixteen shots from the ship's cannons, 'which was the number of years he liv'd in this uncertain world'.[78]

Phillips did not write any more about his feelings. He was a practical man, hardened by years at sea. His journal is full of useful details: wind direction and velocity, moorings, island reliefs, preferred merchandise and prices on different markets, where one can use false weights in trading and where one cannot, which markets sold healthy men and women and how much they cost, and the tricks used by merchants to make their slaves look young and healthy.

In Whidaw, now Benin, Phillips and his business partners bought 1,300 slaves over a period of 9 weeks and divided them up between two ships. The *Hannibal* set sail with 700 shackled human beings who could not stand on the low, lightless and unventilated slave decks, and lay so tightly together that it was impossible to move without stepping on other people.

Phillips was new in the slave business and received good advice from more experienced colleagues, though he did not always take it:

> I have been inform'd that some commanders have cut off the legs or arms of the most wilful, to terrify the rest, for they believe if they lose a member, they cannot return home again: I was advis'd by some of my officers to do the same, but I could not be perswaded to entertain the least thoughts of it, much less to put in practice such barbarity and cruelty to poor creatures,

who, excepting their want of Christianity and true religion (their misfortune more than fault), are as much the work of God's hands, and no doubt as dear to him as ourselves; nor can I imagine why they should be despis'd for their colour, being what they cannot help, and the effect of the climate it has pleas'd God to appoint them. I can't think that there is any intrinsick value in one colour more than another, nor that white is better than black, only we think it so because we are so, and are prone to judge favourably in our own case, as well as the blacks, who in odium of the colour, say, the devil is white, and so paint him.[79]

This is the contradiction of Captain Phillips. He knew what he was doing; he did not consider Africans inferior and did not see himself as belonging to a superior race. For the blacks, the devil is white.

The voyage of the *Hannibal* turned into a financial disaster for the investors. More than half of the slaves died during the journey, either because they caught smallpox or fever, contracted gangrene through their wounds from the chains and lashes, jumped into the sea or, without any given reason, out of despair, which often frustrated Phillips: 'The negroes are so wilful and loth to leave their own country that they have often leap'd out of their canoes, boat and ship, into the sea, and kept under water till they were drowned',[80] he wrote.

For the captain, 370 slaves were a bitter loss, as he stated himself:

there happen'd such sickness and mortality among my poor men and negroes, that of the first we buried 14, and of the last 320, which was a great detriment to our voyage, the royal *African* company losing ten pounds by every slave that died, and the owners of the ship ten pounds ten shillings [...] whereby the loss in all amounted to near 6560 pounds sterling.[81]

Phillips had the misfortune to ply his frightful trade before the caste of explainers and justifiers in his country had turned its attention to slavery. Although he considered his prisoners 'slaves of the devil' because they did not want to hand over their children for missionary tuition on board, his piety and his violence towards people whom he considered equal (aside from the matter of baptism) inhabited different corners of his mind. Nonetheless, the captain only undertook this one voyage. After his return, the money he had made from selling the remaining half of the slaves he had originally loaded was enough to finance a modest but comfortable retirement in his home town in Wales.

The profits from the slave trade were phenomenal. Investors could double or treble their money with a single voyage, since Europe's hunger for cane sugar, coffee, tobacco, tea and other products was insatiable. First it had been white prisoners toiling on the plantations; between 1654 and 1685 alone, 10,000 white convicts were sent from Bristol to the West Indies for forced labour. Soon their work was no longer enough, however, since they could only be used for the duration of their legal punishment and were set free after that. But Africans, who were not Christians, were not subject to these limitations. Christian and non-Christian prisoners worked alongside one another on the plantations.

But then came the missionaries, especially the English Quakers, who insisted passionately on saving the immortal souls of all people, including those whom the Lord had given a dark skin colour. The plantation owners immediately and quite correctly saw this as a threat to their business model and made laws to prevent the release of African slaves who had converted. At the same time, the missionaries also disrupted the social structure of the plantations. Until now, Christians had been preferred workers and the heathens slaves for life; but now there had to be a new dividing line to preserve the distinction between the Europeans, who merely had to work off a temporary punishment, and the slaves.

From now on, the laws of the plantation islands distinguished between white and black labourers. A religious dividing line that made sense from the European perspective of religious wars and confessional states was replaced by a separation according to skin colour. Different skin colours had, of course, been observed and used for discrimination before that, but the lines of discrimination had been different. White Europeans living in primitive conditions could also be 'savages', while black rulers or commanders could be described with great respect, and travel reports emphasized the cultural development of other civilizations, not the colour of their skin.

The enslavement of black people constituted a moral problem, but economically it was almost irresistible. The English poet William Cowper (1731–1800) summed up the dilemma:

I own I am shocked at the purchase of slaves,
And fear those who buy them and sell them are knaves;
What I hear of their hardships, their tortures and groans,
Is almost enough to draw pity from stones.
I pity them greatly, but I must be mum,
For how could we do without sugar and rum?[82]

Even though slavery was an outrage, it soon became impossible to imagine the European market without its products and other useful effects. Slave ships had to be built, equipped, manned and insured; industrially manufactured exchange articles and other important accessories such as chains, leg irons and firearms were produced without pause and boosted the economy; selling slaves made the traders rich, and the products from the plantations were welcomed enthusiastically by the European middle class as status symbols and luxury foods.

The historian Eric Williams published a study in 1944 on the economic significance of slavery for English industrialization that was met with strong agreement from some and outraged rejection by others, for it made two claims: firstly, that the slave trade and its profits substantially enabled the Industrial Revolution, and that England in particular could not have built up its textile industry without this revenue; and secondly and more seriously, that racism was not a prerequisite but a consequence of slavery.

Even if the first claim was controversial, and to an extent still is (retrospective economic calculations are often complex, sometimes contradictory and almost always incomplete), Captain Phillips's journal from 1693 supports the second. Phillips knew the Africans as 'slaves of the devil', but not as inherently inferior human beings.

The journals of later captains paint an entirely different picture. The cruelty of the slave trade and the sadism of captains, seamen, soldiers and plantation owners exceed the imagination. The slaves, who were often packed together for weeks without fresh air and only an hour's exercise per day, lay in their own excrements and fell victim to diseases that spread like wildfire below deck. Many died from diarrhoea. Time and again, seafarers noted that dead slaves were brought up every morning and directly thrown overboard 'like old bottles', as one ship's doctor wrote in disgust. Some of the deported committed suicide, others went mad in their captivity or fell into a deep depression that usually ended fatally.

Brandings, lashings, mutilations, executions and demonstratively cruel punishments to deter the other slaves were normal, especially after mutinies, though God and his grave were never far away, as the following logbook entry shows:

SHIPPED by the grace of God, in good order and well condition'd [...] in and upon the good Ship call'd the MARY BOROUGH, whereof is Master, under God, for this present voyage, Captain David Morton, and now riding at Anchor at the Barr of Senegal, and by God's grace bound for Georgey, in South Carolina, to say, twenty-four prime Slaves, six prime women Slaves, being mark'd and number'd as in the margin, and are to be deliver'd, in the like good order and well condition'd. [83]

The slave traders soon learnt to justify their wealth using the Gospels and the Enlightenment. A certain Captain William Snelgrave already compiled an entire list of good reasons as early as the first half of the eighteenth century. He insisted that blacks did not value life at all and killed one another without hesitation. In addition, the slaves had been enslaved not by whites but by other Africans, quite rightfully, since, according to the laws of their own countries, they were prisoners of war or debtors who could not pay. Others, Snelgrave argued, deliberately produced numerous children in order to sell them as slaves, even though they were not poor. And the Europeans were doing them a favour in any case, since the prisoners of war would simply be murdered or sacrificed to barbarian gods otherwise, and 'when they are carried to the plantations, they generally live much better there, than they ever did in their own country; for as the planters pay a great price for them, 'tis their interest to take care of them'.[84] Granted, it was not always a pretty business, but 'let the worst that can be said of it, it will be found, like all other earthly advantages, tempered with a mixture of good and evil'.[85]

The American doctor George Pinckard visited a slave ship in 1795 to gain an idea of the controversial trade. As he later wrote to a friend, he found a cheerful, friendly and clean ship where the young slaves performed tricks for him while the young women flirted with 'significant gestures' towards him (female slaves were routinely raped by officers during the voyages). However, the young physician also found an opportunity to assure himself of the fundamental otherness of the slaves: 'In dancing they scarcely moved their feet, but threw about their arms and twisted and writhed their bodies into a multitude of disgusting and indecent attitudes. Their song was a wild and savage yell, devoid of all softness and harmony, and loudly chanted in harsh monotony.'[86] He therefore concluded: 'Our minds, necessarily, suffered in contemplating the degrading practices of civilized beings towards the less cultivated heathen of their species; but the eye was not shocked by the abuses of tyranny and inhumanity.'[87]

The young doctor was not the only educated American who reflected on the institution of slavery and evidently did not have an entirely clear conscience. Thomas Jefferson, one of the American founding fathers and owner of 600 slaves, found himself in the strange situation of defending the institution by pointing out that the ancient Romans and Greeks had been far crueller to their white slaves, while also emphasizing that blacks were no less intelligent and moral than whites, but the latter were better able to make use of their reason and ethics because of their position. He strongly cautioned against drawing conclusions about the character and qualities of entire peoples based on a few observations, for 'Our conclusion would degrade an entire race of men.'[88] To Jefferson it is ultimately a question of good science; he does not have information, so he can only formulate his opinion as a hypothesis: 'I advance it therefore as a suspicion only, that the blacks, whether originally a distinct race, or made distinct by time and circumstances, are inferior to the whites in the endowments both of body and mind.'[89] A philosophical mind at work.

Stuffed and Exhibited

A shift in public image and cultural perception also took place far away from the plantations and Atlantic slave ships: in Vienna, where a young African by the name of Angelo Soliman (1721–96) had arrived. There, the discussion about the status of Africans was not as advanced, and probably not as urgent, as in countries directly involved in the trans-atlantic slave trade. Nonetheless, the Habsburg Empire had points of contact with slavery in the Mediterranean through its properties in Italy, and so it came to pass that a young African who was probably born in present-day Niger and abducted from there by human traffickers was taken via Sicily to Vienna, where he worked as a private tutor at the court of the Prince of Liechtenstein.

Angelo Soliman's story has often been told, because it is so typical of a culture's approach to a kind of otherness that was only rarely encountered in German-speaking countries, where few people set eyes on Africans. At the same time, the idea of the African, based on a relatively narrow range of clichés and expectations, was widespread.

As a member of an important prince's court, with direct access to the pro-Enlightenment Emperor Joseph II – who was also his Masonic brother – Soliman was a respected figure in Vienna; he married a Viennese woman and had a child with her. He was not ostracized for the colour of his skin, even though he sometimes appeared at the court in an Ottoman-inspired outfit with a turban – an imaginary national dress also found among Serbs, Hungarians or Turks.

After his death, however, his story took a brutal turn. Franz I, the successor to his unloved uncle Joseph, had the body of this African favourite of a few Viennese nobles exhumed, stuffed and exhibited as a 'savage', dressed in nothing but peacock feathers and glass beads, in his natural history museum, where the exhibit provided visitors with a gruesome thrill for decades before, in an incident of poetic justice, it was destroyed in 1848 by a fire at the court castle. A cultivated man and

13 *Angelo Soliman* (1721–96), tutor at the court of the Prince of Liechtenstein in
Vienna from 1734. Portrait, mezzotint, 1796. Austrian National Library, Vienna

baptized Christian from another continent had become a dangerous
'savage' who, as such, seemed to have forfeited any right to his body, a
decent burial and thus his soul. The change of categories was complete.

Here the justification industry of Western societies was decisive.
John Locke's support for slavery in the USA had been based on naked
economic interests, but in the eighteenth and especially the nineteenth
century, this industry was booming like never before. The subjugation of
other humans could only be morally defended if those humans were not
only contemptible but objectively inferior, closer to nature and animals
than to civilized people: a fundamentally different species. Descartes had

already claimed, against all empirical evidence, that animals had no soul and hence no emotions, no consciousness and no real personality.

In the course of the immensely lucrative triangular trade that brought sugar, cotton, tobacco and coffee to Europe, and in light of the fabulous profits that financed palatial mansions and elegant town houses in England, chateaus in France and elegant villas in many other countries, this technique was now also applied to the people whose unpaid work sustained major parts of the system.

The debate about the division of people into races was still completely open in the eighteenth century. It was the time of the *Encyclopédie* and great classifications, of Carl von Linné's plant systematics and the categorization of animals by the Comte de Buffon and others, of schematizing the earth's layers and chemical elements. These attempts were concerned with a system of nature, but not a hierarchy. Linné did not find any one class of plants more 'valuable' than another, even though the systematization itself was an element in the subjugation and objectification of natural contexts.

The great *Encyclopédie* (published 1751–72) of Diderot and d'Alembert was considered scandalous because it subverted and questioned existing hierarchies of knowledge and society through its alphabetic order. It knew *sauvages* of different skin colour, but no hierarchy of races. Diderot himself even presented the 'Bushmen' in southwest Africa and the indigenous population of Tahiti as superior to Europeans in terms of morality and practical wisdom. He and his radical friends had their collective prejudices too, and sometimes thought in stereotypes, but they did not follow a racist world order. On the contrary: they had most contempt for the church, which they fought wherever possible. Diderot wrote passionately against slavery and colonial subjugation – but censorship forced him to do so anonymously.

Nonetheless, even for the leading philosophical minds of the Enlightenment and the early nineteenth century – and perhaps for them in particular – the division of humans into races increasingly became an article of faith that was repeated incessantly, with no need for further evidence.

Georg Wilhelm Friedrich Hegel (1770–1831) had a great deal to say about Africans and other non-Europeans in his *Philosophy of History*: 'From these various traits it is manifest that want of self-control distinguishes the character of the Negroes. This condition is capable of no development or culture, and as we see them at this day, such have they always been.'[90] Even then, pedantic contemporaries might have countered that reducing all groups of people living on the African continent to a single word to describe not only their appearance, but also their character and their abilities, seemed rather a generalization; but Hegel had no such scruples, and his followers repeated his words like Bible verses.

Hegel, who never left Germany except for a brief stint as a private tutor in Bern, wanted his thinking to encompass the world. To him, history was the progression of the world spirit towards the realization of freedom, and whoever seemed to reject this progress or, in his view, was unfit for it had lost every civil right. One can hear the voice of the slave captain William Snelgrave in the background of the German thinker's arguments: the Africans know nothing else, they deserve nothing else, they are incapable of anything else; their new owners are doing these poor savages a favour. This is how Hegel puts it:

Negroes are enslaved by Europeans and sold to America. Bad as this may be, their lot in their own land is even worse, since there a slavery quite as absolute exists; for it is the essential principle of slavery, that man has not yet attained a consciousness of his freedom, and consequently sinks down to a mere Thing – an object of no value. Among the Negroes moral sentiments are quite weak, or more strictly speaking, non-existent. Parents sell their children, and conversely children their parents, as either has the opportunity. Through the pervading influence of slavery all those bonds of moral regard which we cherish towards each other disappear, and it does not occur to the Negro mind to expect from others what we are enabled to claim. […] In the contempt of humanity displayed by the Negroes, it is not so much a despising of death as a want of regard for life that forms the characteristic feature.[91]

No cliché is too pathetic for the great man to repeat with all the authority of a royal Prussian professor:

The Negroes indulge, therefore, that perfect *contempt* for humanity, which in its bearing on Justice and Morality is the fundamental characteristic of the race. They have moreover no knowledge of the immortality of the soul, although spectres are supposed to appear. The undervaluing of humanity among them reaches an incredible degree of intensity. Tyranny is regarded as no wrong, and cannibalism is looked upon as quite customary and proper. Among us instinct deters from it, if we can speak of instinct at all as appertaining to man. But with the Negro this is not the case, and the devouring of human flesh is altogether consonant with the general principles of the African race; to the sensual Negro, human flesh is but an object of sense – mere flesh.[92]

Faced with such inferior creatures, clearly quite incapable of rational thought and moral behaviour, it is only logical that Hegel considered slavery more an instrument of the world spirit than a moral scandal. Captains had unanimously reported that their prisoners were so fearful of the voyage because they were convinced the Europeans would kill and eat them or burn their bodies to coal. 'Who is the barbarian?', Montaigne would have asked.

Hegel also reflected on the settlement of the Americas: America was 'the land of the future', he recognized, 'a land of desire for all those who are weary of the historical lumber-room of old Europe'.[93] Those people were brave spirits who wished to leave a now-dusty continent and head towards a glorious new tomorrow, especially since the new continent displayed a 'physical immaturity' and was still almost entirely unexplored; those who carried out the advancing will of the world spirit were always in the right. Before the 'discovery' of America, there was barely any culture there worth mentioning, Hegel writes, for the 'natives' were characterized by a 'mild and passionless disposition, want of spirit, and a crouching submissiveness' until the Europeans 'succeed in producing any independence of feeling in them'. He continues: 'The inferiority of these individuals in all respects, even in regard to size, is very manifest.'[94]

It was clear, then, that such inferior individuals could have no claim to this majestic land of the future, but, rather, the 'industrious Europeans, who betook themselves to agriculture, tobacco and cotton-planting etc.'.[95] The 'natives' were not only incapable of living in a civilized fashion; in South America, they were so lazy that 'at midnight a bell had to remind them even of their matrimonial duties'.[96] Nonetheless, there was great work to do there, and this suggested taking a logical step: 'The weakness of the American physique was a chief reason for bringing the Negroes to America, to employ their labour in the work that had to be done.'[97]

According to Hegel, the 'elevation of man over Nature' only consists in his superior mental powers, which are most fully manifest in Europeans and not fully developed, or perhaps already withered, in Asia. Only through his spirit can he see that 'the chance volition of man is superior to the merely natural – that he looks upon this as an instrument to which he does not pay the compliment of treating it in a way conditioned by itself, but which he commands'.[98]

There is a straight line leading from Hegel's armchair rhetoric to the constitutional scholar Carl Schmitt (1888–1985), who was not only a staunch National Socialist but also one of the most influential legal philosophers and clear-sighted theorists of power in his generation. 'Land appropriation' on other continents, Schmitt writes, was entirely legitimate and did not change European law. Diderot had already asked what would happen if a Tahitian went to a French beach and staked a claim to the entire land; after all, if land appropriation is lawful, it must be so in every case. Schmitt does not accept this argument, however:

> This soil was free to be occupied, as long as it did not belong to a state in the sense of internal European interstate law. The power of indigenous chieftains over completely uncivilized peoples was not considered to be in the public sphere; native use of the soil was not considered to be private property. [...] Whether or not the natives' existing relations to the soil – in agriculture, herding, or hunting – were understood by them as *property* was an issue to be decided by the land-appropriating state. [...] Just as in international law the land-appropriating state could treat the public property (*imperium*) of appropriated colonial territory as leaderless, so it could treat private property (*dominium*) as leaderless.[99]

Hegel and Schmitt were separated by a century and two world wars, but the arguments remained in place. Generations of thinkers had contributed to this: Immanuel Kant, the moral patron saint of the Enlightenment and German Idealism, whose world was restricted to Königsberg and who, through his friendship with English traders in the city, at least had an indirect connection to the slave trade, held very clear views about Africans. There is a note indicating that he met a dark-skinned person once, but the encounter was clearly not marked by mutual respect. As a young man, in 1764, Kant wrote the following note after a discussion about art: 'this scoundrel was completely black from head to foot, a distinct proof that what he said was stupid'.[100] Decades later, he wrote: 'The Negro can be disciplined and cultivated, but is never genuinely civilized. He falls of his own accord into savagery. Americans and Blacks cannot govern themselves. They thus serve only for slaves.'[101] But the Africans did not mark the nadir of humanity, Kant argued. The indigenous peoples of America, as Hegel agreed, were too lazy and worthless even to survive by themselves: 'this race, too weak for hard labour, too phlegmatic for diligence, and unfit for any culture, still stands – despite the proximity of example and ample encouragement – far below the Negro, who undoubtedly holds the lowest of all remaining levels by which we designate the different races'.[102]

If I have only cited German authors here, this is partly because it shows how far the division of humans into races that grew from the justification of the transatlantic enslavement of Africans had penetrated societies that were only indirectly, or only to a lesser extent, involved in slavery. Among the Enlightenment thinkers alone, however, a Montesquieu or Hume expressed very similar views, while Voltaire, that apostle of tolerance and reason, returned to the subject time and again: 'I see men who seem to me superior to these negroes, as the negroes are to the apes, and as the apes are to oysters.'[103]

The Silent Death of Saartjie Baartman

The subjugation of nature reached a new level through the moderate Enlightenment, industrialization, the systematic exploitation of colonies and the incipient natural sciences, but it still carried its old moral-theological burden. It had to be proved that subjugation was not only possible and useful for the subjugators, but actually good and just – God's will, or the will of providence, of the world spirit, of progress, not despotic power and cruelty.

Just as subjugation had been theologically underpinned in earlier centuries, it was increasingly supported over the course of decades by scientific methods and modern technologies. Carl Hagenbeck's dire human zoos were promoted as being educational for the people, but their spectacular performances transported the carefully constructed stories that declared so-called 'savages' a part of the nature that was to be subjugated. For this pretty scenery, which was literally fabricated in the case of the ethnographic exhibitions, facts were an obstacle and rational thought was completely dispensable. What facts did the philosophers cite to support their claims? How plausible was it that there was just one homogeneous 'race' living on each of the enormous land masses of Africa and the Americas, one that was incapable of any genuine culture? Had they never picked up the travel reports of Georg Forster or Alexander von Humboldt, or the oft-translated ones by Bougainville or the accounts of Jesuits from South America? Each of these widely known and cited sources painted a more nuanced, often highly respectful picture of the societies encountered by the Europeans. Nonetheless, the intellectual pioneers of subjugation were able to make their *ex cathedra* pronouncements with absolute conviction, uttering truths borrowed from the discourses of slavers, and still be taken seriously by their readers. Beside Nietzsche's slave morality, which we will encounter later on, there was always a slaver morality too.

In the nineteenth century, the idea of subjugation, of natural hierarchies with white Europeans as the apex and culmination of all known life forms and the existence of human races, had solidified into a scientific truth. The German naturalist Ernst Haeckel occupied himself not only with his almost pantheistic idea of art forms in nature, which is still aesthetically astonishing today, but also with series of skulls showing the descent of Europeans from apes via other ethnicities. In Italy, the criminologist Cesare Lombroso worked on a strict scientific inventory of all physiological traits for a quantitative measurement of criminal tendencies, ethnic differences and human value.

The eminent French anatomist and palaeontologist Georges Cuvier measured skulls and brain volume to prove his theory that different human races were all created from a white Adam, and that the darker forms were degenerations of the original light-skinned perfection. Based on their appearance, he too claimed that Africans were related to hordes of apes and had remained in 'the most complete state of utter barbarism'.[104]

This attitude made it easier for Cuvier to carry out one of his most famous studies. The object of his rigorous scientific curiosity was a certain Saartjie Baartman, a young woman who was born in South Africa and taken against her will to France. Because of her large bottom, she was exhibited for money at fairgrounds in England, then in the Netherlands and France. She was also hired out as an exotic prostitute, first to wealthy clients and then to more ordinary ones.

Cuvier was not interested in the sad fate of the young woman, who had by that time been destroyed by syphilis, solitude and alcoholism and was living in abject poverty. He saw her as the confirmation of his ideas about human races. He measured her carefully and had her painted nude shortly before her death at the age of twenty-five. Then he dissected her body in the service of science, paying particular attention to her brain and genital organs. Her skeleton could be viewed at the Musée de l'homme in Paris until the late twentieth century.

The reference to hordes of apes inevitably leads us to Charles Darwin (1809–82 – another scientist who was convinced that Africans are closer to apes than to humans), whose unintended intellectual revolution through the publication of *On the Origin of Species* (1859) inflicted one of the central narcissistic injuries in the history of the Western subjugation of nature. From the Neoplatonists to the moderate Enlightenment philosophers, generations of preachers with all sorts of messages had instilled the exceptional status first of humans, then of white humans, into the minds of those who followed that history. Darwin's theory of evolution had an undeniable explanatory power, but it precluded that exceptional status.

As a leading scientist, Cuvier had still attempted to reconcile the creation story with palaeontological finds and to trace all humans that had ever lived back to a physically existing Adam created by God. The logic of evolution, however, finally made it impossible to conceive of a human being that had remained unchanged since the creation roughly 6,000 years ago. The law of the development of species through adaptation to changing conditions knew no exceptions and stipulated that humans and other primates must have had shared ancestors, and that the difference between them could only be a gradual one. One must acknowledge, Darwin wrote,

> that man with all his noble qualities, with sympathy which feels for the most debased, with benevolence which extends not only to other men but to the humblest living creature, with his god-like intellect which has penetrated into the movements and constitution of the solar system – with all these exalted powers – Man still bears in his bodily frame the indelible stamp of his lowly origin.[105]

Darwin cast man from the throne on which the Bible had placed him. He was not especially happy with this, since he was in poor health and feared terrible trouble with friends and acquaintances, to say nothing of church leaders. This assessment proved accurate, for the publication of his theory, which he himself had delayed several times, caused a storm of indignation that cost him a great deal of time and other energy that he would rather have devoted to his lifelong passion for insects or other experiments, such as the behaviour of seeds in salt water or the sense of hearing in earthworms.

The idea of subjugation could not be done away with by a single theory, however; on the contrary, it adapted. Of course, there was (and is) still a chorus of voices that refused to accept the position of humans as natural beings and continued to insist on their divine origin and unique, incomparable character, but they were fighting a losing battle in the scientific world.

Although it could no longer invoke the special biological status of humans, the subjugation of nature nonetheless became part of evolutionary thought, especially thanks to Darwin's distant cousin Sir Francis Galton (1822–1911), a scientist whose wide-ranging intellectual interests, immense energy and excellent connections made him a leading authority in such diverse fields as fingerprint research, meteorology, statistics and African studies. He championed selective reproduction among (white) humans for an improvement of the future.

Galton was an astonishing man who had inherited a certain moral schizophrenia. His family had made its money in the arms trade, but also belonged to the pacifist Protestant community of Quakers. His birth house had been built for Joseph Priestley, incidentally – the same scientific genius we encountered in our consideration of the vacuum pump in the painting by Joseph Wright of Derby. Scientific England was small. Galton studied law and mathematics, travelled through South Africa, wrote on all manner of subjects that relied on data and statistical methods and pondered the future of humanity. In a letter to *The Times*, he expressed the view that the Chinese should be encouraged to immigrate to Africa so that, thanks to their proverbial diligence, the land there could finally be cultivated and systematically exploited, and the obviously inferior blacks replaced.

It was only when Galton read *On the Origin of Species* that the scientific universalist found his true calling. From that point on, he devoted himself entirely to eugenics, the creation of a race of improved humans through selective reproduction. It was as much a political as a biological programme, and Galton gained admirers such as Winston Churchill and George Bernard Shaw. His own ideas about the form of superiority were characterized by a mixture of race theory and social Darwinism. On his trip to Africa he had already noted:

> The greater part of the Hottentots about me had that peculiar set of features which is so characteristic of bad characters in England, and so general among prisoners that it is usually, I believe, known by the name of the 'felon face'; I mean that they have prominent cheekbones, bullet-shaped head, cowering but restless eyes, and heavy sensual lips, and added to this a shackling dress and manner.[106]

The faces of the indigenous Africans reminded Galton of Cesare Lombroso's photographic galleries of delinquents, imbeciles and violent criminals. Both ethnic deficiencies and those of character or intellect were to be eliminated through selective reproduction.

Galton was undoubtedly a brilliant mind, but great intelligence shields neither from prejudice nor from extreme stupidity in other matters. When his studies of ethnic types led him to the children at a Jewish school in London, he wrote the following in his report on his experiments: 'The feature that struck me most as I drove through the adjacent Jewish quarter, was the cold, scanning gaze of man, woman, and child.'[107] For this gentleman from one of the wealthier parts of town, such cold calculation could only mean one thing: 'I felt, rightly or wrongly, that every one of them was coolly appraising me at market value, without the slightest interest of any other kind.'[108]

Galton sought to construct ethnic ideal types through photographic composites (the superimposition of several portraits to form a 'ghost portrait') and was especially enthusiastic about the portraits of Jewish children, which he saw as representing an ideal Jewish type. Jewish colleagues who worked on the study with Galton had an entirely different interpretation, as they explained to the Anthropological Society in London. Firstly, they argued that European Jews should not be considered a race, since they had become so ethnically mixed over the course of their history that they were no different from other Europeans; and secondly, their 'cold' facial expressions could be explained by the fact that they were poor, hungry and oppressed. They analysed as a social phenomenon what Galton saw as a biological fact.

Sir Francis Galton's eugenics was especially intended to pass on positive intellectual qualities. He carried out studies of geniuses and artists and quantified his results very meticulously, but could only find weak statistical correlations in families. Ultimately, his ideas were little more than the typical prejudices of an Englishman of his class and generation converted into numbers. He thought that the greater 'simplicity' of black people could even be identified in their fingerprints, which he found more simply structured; he saw the anatomy of criminals in the black body and vice versa; and he considered Jews cold and money-hungry. Despite these obvious flaws, his theory was enormously influential: there were eugenics societies all over Europe and the United States that put on lecture series and international conferences, and his ideas were disseminated via bestselling books and articles in the press.

Perhaps the hidden power of Galton's ideas lay in the fact that he combined Darwin's theory, which was difficult to understand for many, with a newly proven biological superiority of white Europeans and thus restored the familiar order of the world. Even if, in Darwin's formulation, man could not deny his lowly origin, the biological mechanism of selective reproduction could be used to elevate him above all of nature and even to reshape the bodies of humans based on his own seemingly rational criteria.

The hierarchy of races and life forms in general became the most successful way of subjugating nature because it made it possible to reinvent and morally upgrade political and economic power as a law of nature, providence or progress. Whatever belonged to wild nature had to be dominated and improved by civilized humans – and they were the ones who decided the criteria for what was wild and what was not. It was a watertight system of argumentation that could even be reconciled with ideas such as 'liberty, equality, fraternity' as long as one could decide who and what was entitled to freedom, equality and solidarity. Whoever was considered unworthy was a slave, an object under the control of a self-appointed virtuous power.

Equality was mitigated by being turned into equality among those who were alike, and this likeness could be postulated and dissected in ever new ways. No – as well as other 'races' and women, workers and 'common folk' could not be considered true equals either, for they were hereditarily degenerate and barely human, almost sliding back down the ladder of evolution. Such was the view among nineteenth-century scientists as they gazed disdainfully on the industrial proletariat, which they wrote about with the same revulsion as they did about the indigenous peoples of other continents.

Victorian social pioneers described the inhabitants of London's East End like an inferior, barely human race slipping down the evolutionary ladder on the way to becoming animals again. Most of these unfortunates were English or Irish. It was in everyone's best interests to keep power in the hands of those who could administer and defend the ideal of equality in the Athenian sense – as equality between free, wealthy European men. This made it possible for a complete travesty of all the core ideas of Christianity and the Enlightenment to be carried out and officially celebrated by Christian and enlightened people who were convinced of their own righteousness.

The sciences of justification serving the Western compulsion to dominate covered a wide range of fields. They extended from theology and philosophy to new sciences such as anthropology and zoology, from children's books and gummed pictures of fearless explorers via adventures among savages to breathless accounts of glorious battles in the colonies, scientific progress and the grateful smiles of African orphans in the press.

Thinking in Europe during the eighteenth and nineteenth centuries meant living in a world that had lost stable reference points: first in the dispute between philosophy and theology, then with ever new discoveries and inventions, which increasingly called the old order into question. Machines and factories were political events that motivated decisions and created infrastructure; a working class and the mechanisms for its surveillance; strikes and revolutions; radical tracts and sentimental novelettes.

Not all contemporaries welcomed so much upheaval and renewal. Rousseau led the colourful army of those for whom civilization itself was the problem, and in Europe, especially after the French Revolution, a solid phalanx of nostalgic, royalist and often reactionary thinkers formed. They remained a minority, however; it was the propagandists of the new who stood in the limelight. The optimism of some authors was boundless, for they finally seemed to have discovered a way for humanity to free itself from the shackles of nature and the scourges of fate, and to take control of its own destiny.

The idea of subjugation was so tempting and dazzling in its many facets that numerous intellectuals, and even more simple citizens, succumbed to it as if to a welcome infection. From Hegel to Haeckel, from the slave ship *Hannibal* to Hagenbeck's ethnographic exhibitions, from natural science to anthroposophy, from toys to soapbox sermons, history and geology or theology and novelettes, the most varied genres celebrated the dream of absolute subjugation and its optimistic twin, progress. God was on their side.

The British publication *The Pulpit Commentary*, the 48-volume compendium for church sermons in Victorian England, traces the apotheosis of human domination to the Bible and encourages more appropriation of land. It suggested the following for a sermon on Genesis 1:28 ('fill the earth and subdue it'):

> At the present day man has wandered to the ends of the earth. Yet vast realms lie unexplored, waiting his arrival. This clause may be described as *the colonist's charter*. **And subdue it.** The commission thus received was to utilize for his necessities the vast resources of the earth, by agricultural and mining operations, by geographical research, scientific discovery, and mechanical invention. **And have dominion over the fish of the sea**, &c. i.e. over the inhabitants of all the elements. The Divine intention with regard to his creation was thus minutely fulfilled by his investiture with supremacy over all the other works of the Divine hand.[109]

This 'man' who was described as wandering was not only implicitly male, but also implicitly white; he found the rest of the Earth untouched, in a sense virginal, for the unproductive savages living there had forfeited their right to property through their inability to do anything useful with the gifts of the Lord. It is therefore a direct mission from the Lord that one should clear forests, cultivate fields, dig mines in the ground, make measurements and inventions in order to perfect the domination of nature.

Thus the divine hand guided every lucrative undertaking; every theft was commanded from on high. The scruples of a Captain Phillips, who saw no difference between himself and the 'poor creatures' he (nonetheless) transported, had long been theorized away. The people in the colonies were, in the immortal words of Rudyard Kipling, no more than 'new-caught, sullen peoples, / Half-devil and half-child'. Whoever stood in the way of progress also stood in the way of the Lord's will or, for those of a secular cast, the world spirit.

With refreshing clarity, the British-Indian historian Pankaj Mishra dissects progress from a Western perspective, and the selective blindness that accompanies it, for this progress always leaves a trail of destruction in its wake. The perspective of those whose lives have been destroyed, however, is rarely part of the memory. Be it deforested areas, drained soil, overfished oceans, extinct animals, destroyed habitats, human ways of life and identities eliminated, especially in the course of the Industrial Revolution, with the resulting immiseration or fear thereof, and then the reactions to this, extending to pogroms and wars – the beneficent power that produces more and more commodities, services and scientific discoveries always exacts a price. 'Obscuring the costs of the West's own "progress", it turns out, severely undermined the possibility of explaining the proliferation of a politics of violence and hysteria in the world today, let alone finding a way to contain it.'[110]

*

Progress claimed victims. With its immense dynamics, it devastated traditional ways of life, economic practices and political structures, mercilessly putting its stamp on them before formulating an entire bouquet of reasons why this was the only way, the right way and the moral way. But the old structures that were erased by this destructive wave of the spirit left behind ugly, gaping holes, along with a history of traumatization and humiliation whose effects might still outflank the history of progress and growing prosperity.

The old structures of work and society, the old hierarchies and certainties, were unsettled by the attack of the Industrial Revolution and a new subjugation by the logic of imperialist domination, the power of nationalist ideas and economic necessity. The transcendence of this new domination was progress itself, the expansion of the sphere of power and the fulfilment of a historical mission.

One idea in particular seemed less and less central in the context of science, as the charismatic general Napoleon Bonaparte discovered. The French mathematician and astronomer Pierre-Simon Laplace (1749–1827) was a brilliant scientific pioneer who made a living as an examiner at French artillery schools; at one of these, he had examined a short sixteen-year-old cadet named Napoleon Bonaparte. Years later, when Napoleon was a successful general and First Consul of France and his erstwhile examiner an eminent astronomer, Laplace handed his former examinee his latest book *Traité de mécanique céleste* (Treatise on Celestial Mechanics).

The young general, who was still short, clearly relished making his teacher feel his newly won power and rebuked him in front of everyone present. Newton had spoken of God in his great work, Bonaparte criticized, but he had not found a single reference to God in this book.

'Sir', Laplace replied with icy cordiality, 'I have no need of that hypothesis.'[111]

Hare Hunting

God is dead! God remains dead! And we have killed him! How can we console ourselves, the murderers of all murderers! The holiest and the mightiest thing the world has ever possessed has bled to death under our knives: who will wipe this blood from us? With what water could we clean ourselves? What festivals of atonement, what holy games will we have to invent for ourselves? Is the magnitude of this deed not too great for us? Do we not ourselves have to become gods merely to appear worthy of it? There was never a greater deed – and whoever is born after us will on account of this deed belong to a higher history than all history up to now!

Friedrich Nietzsche, *The Gay Science*[112]

It was progress itself that had claimed this last victim. In a universe that could be described with mathematically expressed laws, there was simply less and less room for a transcendental relation, an immanent divinity. Here, at the zenith of its trajectory, the idea of subjugation was confronted with its inner contradiction. At some point, the humans who had forced nature to its knees by order of God no longer needed the old man. In so far as God was reduced to the great watchmaker, however, liberated humanity had to shoulder the burden of responsibility for its own actions.

If the world could be explained on its own terms and humans were no longer acting on a historic mission – who were they then? And who could answer that question for them? Where could they flee from this emptiness? And who could save them from the guilt of the mighty?

At the official and institutional level, such dangerous questions did not lead to a collapse of religious thought, but to an even more intense and dogmatic campaign of justification in sermons, editorials, lectures and edifying plays. The Victorian bourgeoisie needed its divine mission as urgently as the interest from its stocks in the Indian Railway, and the two were disastrously connected. Without a moral substructure, without the 'white man's burden', without a *mission civilisatrice*, there was no colonial empire and no income. The debate about the slave trade and its elevation to a noble sacrifice by white benefactors had shown that centuries of biblical exegesis had borne fruit, and that any obvious fact could be inverted by skilled and well-trained interpreters.

What repeatedly became clear in the debates about the Enlightenment was the obvious need, and even personal desperation of some participants, to cling to the intimate transcendental relation to a ground of all morality and truth, without simultaneously having to reject the obvious advances of science and their problematic explanations. The system needed to preserve God in some form, even if it was only as a watchmaker, a principle, as the highest reason.

The battle for God was no less than a spiritual avalanche. It is difficult to imagine today how intimate and existential was the premodern relationship of many (if not all) European Christians to their God. The physical world alone was pervaded by secrets and miracles. Whatever could not be seen with the naked eye was unknown, and heaven was a tent stretched around the Earth before the first microscopes and telescopes opened up new worlds around 1600. But even they and their insights were only accessible to a select few, and the scientific method was still in its infancy. God's truth provided an explanation that was no less plausible than others. Life seemed full of insoluble mysteries, and it was only through religion that they acquired a visual language, became imaginable and could be resolved in that language. At the same time, the Bible spoke of a world that was familiar to the rural population, since it dealt with farmers and their lives.

However, this explanation and the resolution of all contradictions in the wonderfully paradoxical Holy Trinity were stymied by the fact that the world outside of the religious mindset – the world of natural laws, countries and civilizations on previously unknown continents, the knowledge of comparative anatomy and contemporary cultural anthropology, archaeology and geology – created an opposing image with the decisive advantage that one could use it to navigate the globe, do lucrative business and build vacuum pumps. But this scientific antithesis had a decisive psychological disadvantage: unlike the revealed truth of religion, the knowledge of experiments and theories was only ever a mere model, doomed to be falsified and replaced at some point. It lacked the dignity of a timeless revelation or the glory of a vision, but its prosaic correctness was more helpful than any Bible verse when it came to calculating bridge arches and launch angles.

The enlightened battle over God and his secular equivalents, such as progress and destiny, raged with great intensity in the course of the nineteenth century, for the energies driving the rapid development of industries, colonial empires, large cities and mass culture, technologies and theories about the natural world could no longer be controlled, and shattered all traditional certainties. Colonized peoples and European day labourers, children in factories, landless farmers and people fleeing from hunger and misery had to adapt to the new circumstances – or starve. The effort involved in compensating for this by presenting these processes as necessary, even virtuous and right, was immense, but there were enough minds willing to take on the task. The violence that Pankaj Mishra locates in this process was hidden by increasingly artfully decorated facades.

But these facades also provoked internal contradictions and aroused curiosity. Sigmund Freud grappled with the psychological facades of the Viennese bourgeoisie and found a chaos of conflicting emotions behind them whose effects were too destructive for their existence even to be acknowledged. In *Civilization and Its Discontents*, the founder of psychoanalysis pondered how such aggressive and destructive impulses could be tamed by turning them inwards:

> His aggressiveness is introjected, internalized; it is, in point of fact, sent back where it came from – that is, it is directed towards his own ego. There it is taken over by a part of the ego, which sets itself over against the rest of the ego as a superego, and which now, in the form of 'conscience', is ready to put into action against the ego the same harsh aggressiveness that the ego would have liked to satisfy upon other, extraneous individuals. The tension between the harsh superego and the ego that is subjected to it, is called by us the sense of guilt; it expresses itself as a need for punishment.[113]

This superego, Freud argues, functions like 'an occupation in a conquered city' and rules via the conscience. He soon brings up the sense of guilt, which can even be awakened by a mere intention or desire; he views this as the internalization of a real guilt far in the past

– namely, that of the son who murders his beloved and hated father in order to replace him.

> After their hatred had been satisfied by their act of aggression, their love came to the fore in their remorse for the deed. It set up the superego by identification with the father; it gave that agency the father's power, as though as a punishment for the deed of aggression they had carried out against him, and it created the restrictions which were intended to prevent a repetition of the deed.[114]

An occupation in a conquered city – for centuries, this was also a response to the ideas of a Doctor of the Church, Saint Augustine, who had brought this occupation into the conquered city of the human psyche and, one might say, built a garrison for it there. For Augustine, desire and instinct were obstacles on the path to salvation that had to be controlled and eliminated, by force if necessary, in order to evade hell. Freud countered that the tyranny of the superego could already constitute a form of hell, since desire did not vanish when it was suppressed and denied; it simply began to fester and proliferate below the surface until it had poisoned the whole psyche.

As expected, Freud's interpretation is tailored to the Oedipus complex he had postulated, as well as containing ambivalent emotions. What emerge with Laplace, however, who no longer needed God as a hypothesis, or with Nietzsche, who concluded with dismay that there is no consolation for God's murderers, are the outlines of a collective sensibility among the group of people who had profited most extensively and dramatically from the subjugation of the world and nature: the bourgeoisie, the middle class, the petty bourgeoisie and new city-dwellers, the emigrants and immigrants – in short, the more or less successful and hopeful inhabitants of a still-nascent world of dominance.

The guilt of the new, secular individual haunted the culture of the nineteenth century, a shadow of the old Christian guilt. It drove the heroes and heroines of Verdi's operas and Dostoyevsky's novels, inspired Victor Hugo and Charles Dickens, Goethe's *Faust* and Mary Shelley's *Frankenstein*, kept Kierkegaard awake for nights on end and crushed the unfortunate Woyzeck in Georg Büchner's play of the same name.

Friedrich Nietzsche (1844–1900), born the son of a strict Protestant vicar, was very familiar with guilt and the hypocrisy of Christian morality, which had driven him first into the arms of the far more sensual classical antiquity and then into the rapture of intellectual boundlessness. In his texts, which were poetic, polemical, provocative, paradoxical and philosophical in the truest sense, he railed against the whole of Western culture with its slave mentality and against the moral dwarfing of humanity perpetrated by Christianity. Christian morality and Germany were the two opposing poles of his ambivalence, and they never let him go.

However, even a thinker as experimental as Nietzsche was aware that 'even we knowers of today, we godless anti-metaphysicians, still take *our* fire, too, from the flame lit by the thousand-year-old faith, the Christian faith which was also Plato's faith, that God is truth; that truth is divine'.[115] An important part of this unbroken, albeit often unrecognized, faith is morality, which provides affirmation time and again: '*Morality dresses up the European* – let's admit it! – into something nobler, grander, goodlier, something "divine".'[116] But so much false beauty could not last:

> We Europeans confront a world of tremendous ruins. A few things are still towering; much looks decayed and uncanny, while most things are already lying on the ground: picturesque enough – where has one ever seen more beautiful ruins? – and overgrown by large and small weeds. The Church is this city of decline: we see the religious society of Christianity shaken to its lowest foundations; faith in God has collapsed; faith in the Christian-ascetic ideal is still fighting its last battle.[117]

To Nietzsche, the present of the late nineteenth century was an intolerable compromise between what had been and what was to come, and its inhabitants were homeless in more ways than one:

> We children of the future – how could we be at home in this today! We are unfavourably disposed towards all ideals that might make one feel at home in this fragile, broken time of transition; as for its 'realities', we don't believe they are *lasting*. The ice that still supports people today has already grown very thin; the wind that brings a thaw is blowing; we ourselves, we homeless ones, are something that breaks up the ice and other all too thin 'realities'.[118]

The thin ice of realities was breaking everywhere at a time when the mechanical revolution was devouring millions of people's lives. Especially in Britain and its colonies, the iron fist of industrialization seized society with incredible force. Indian cotton was spun and woven in northern England before being exported again, including to India, but this time at the end of the value creation chain. Through the mechanization of spinning wheels and looms, as well as the use of steam engines powered by coke, this industry, along with the transatlantic triangular trade with slaves, sugar and other products, became a central driver of industrial, social and political development.

The world power Great Britain, which had been no more than a restless island kingdom at the northern end of the inhabitable world only two centuries earlier, now seemed to be a model example of how subjugation in the name of the Lord and progress was possible. Nowhere else was there such an immense, obscene, unprecedented growth in wealth, knowledge, opportunities, theories, trinkets, suburbs, railway lines, novels, luxury boutiques, inventions, edifying sermons and cathedral-like train stations, steel bridges and reports from distant lands – it was a ride in a fast machine that, as church hymns and history books emphasized, was willed and blessed by God.

This overly great unanimity of public opinion in the midst of the unheard-of immiseration of those who generated this wealth, described with the newly coined term 'lumpenproletariat', produced resistance. Politically, this was the moment of different movements inspired by the collaboration between the German exiles Karl Marx and Friedrich Engels. Marx and Engels found enough illustrative material for their studies amid the industrial misery and wealth from London to Newcastle. Culturally, however, the reaction to such apocalyptic transformation and the confrontation of its darker side took place mostly in the milieu of the upcoming middle class: people who possessed enough to learn reading and writing, and to have the necessary leisure to read, yet had not become blind to the ills of society.

One of the central texts of English literature, which already thematized the terrific cost of a purely instrumental relationship with nature at the start of the century, was 'The Rime of the Ancient Mariner' by Samuel Taylor Coleridge (1772–1834), a poem whose folksy tone belies its great sophistication. It can be read as a ghost story, but for many readers it has an almost prophetic quality despite its apparent naivety.

At a village wedding, an old seaman tells the story of a voyage that was cursed. He himself had brought the disaster on the ship when, out of sheer boredom and baseness, he shot an albatross that had been the ship's loyal companion. The weather changed immediately, and his fellow sailors hung the dead bird around his neck as a punishment.

It is only after everyone on board has died that the mariner is illuminated and redeemed, whereupon the crew comes back to life, a merciful wind carries them home and he understands: 'He prayeth best, who loveth best / All things both great and small; / For the dear God who loveth us / He made and loveth all.'[119]

Perhaps it was the harmless air of these lines that made them so powerful. The albatross around one's neck has remained in the English vocabulary as a symbol of a curse and moral burden. The image of a man who, under inhuman conditions, himself becomes inhuman and part of nature, then wilfully destroys a relationship with another creature, was etched into the collective consciousness as much as Dr Frankenstein, the pitiless demiurge who made nature monstrous, or the death-obsessed Captain Ahab sent on his epic final voyage by Herman Melville in *Moby-Dick*, and later the demonic Kurtz in Joseph Conrad's *Heart of Darkness*.

The moment had now arrived in the relationship between humans and nature when humans, not nature, were perceived as the threat. Even the shackled Prometheus was freed from his rock in a poem by Percy Bysshe Shelley (Mary Shelley's husband). Mythology and Olympus itself were no longer safe from human hubris.

Two artists formulated visions that contrasted strongly with the dominant mentality of their time. William Blake (1757–1827) was an engraver, painter, poet and illustrator who had taken non-conformist positions on religion and politics early on. In his graphic works, he spoke out against slavery and created images of human suffering that are unforgettable in their almost naive intensity. In another work he portrayed Newton, the honoured founding father of the scientific revolution. Here it seems to be Newton's heroic spirit rather than his academic persona that is using the compass, for the body is not that of a scholar with long hair, but that of a muscular demigod whose attention is focused completely on his calculations (figure 14).

There is good reason to see a critical element in this representation of scientific thought. The rock on which the scholar's figure sits is overgrown with coral and barnacles, and the whole scene seems to be taking place underwater. But the mathematician is too engrossed in his ideas to notice the abundantly rich nature around him that he wants to describe and comprehend.

Blake himself, whose dream paintings and mythological meditations were often based on visionary experiences, had very different hopes for the material world and the unity of all things: 'To see the world in a grain of sand / And a heaven in a wild flower; / Hold infinity in the palm of your hand / And eternity in an hour.'[120]

We find a second artistic expression of this watershed in the work of Joseph Mallord William Turner (1775–1851), whose storms of colour and quasi-abstract landscapes went in a highly idiosyncratic direction that was far ahead of his own century. Turner too had an unambiguous political position, conveyed in such works as the painting *The Slave*

14 William Blake, *Newton*, 1795. Colour monotype, watercolour, 46 × 60 cm. Tate Britain, London

Ship (originally entitled *Slavers Throwing overboard the Dead and Dying – Typhon coming on*) in a scene he had read about in the newspaper. In order to claim the insurance premium for their cargo, the crew of a slave ship had thrown slaves who had almost died of thirst overboard. The crime became known and resulted in a sensational lawsuit against the captain. In his stormy seascape, Turner shows the bodies of the moribund slaves while the ship is already being assailed by the waves of the typhoon. An ominously blazing sun, surrounded by blood-red storm clouds, hangs threateningly over the entire scene.

Turner chose another maritime scene for *The Fighting Temeraire*, though this time the sea's surface is almost as smooth as a mirror. The once-mighty warship, which helped to create and defend the British Empire, is on her last voyage to be broken up. Her power has faded, her cannons and seaworthy burdens have disappeared and she is being towed by a stout little steamer. All of this mirrors an old time, now defenceless, being forced to follow the new power; one could almost see it as a swan song for the great times of Nelson's fleet and the zenith of global power.

The painting that, in retrospect, is perhaps Turner's most prophetic, shows the spirit of a hare running for its life. The first level is superficial in a dual sense: Turner only added the animal shortly before presenting the canvas in public, and the thin layer of colour has faded over almost two centuries. But the hare was there, and its spirit is still rushing ahead of the train, roughly halfway over the bridge. *Rain, Steam and Speed – The Great Western Railway* (figure 15) was presented to the public in 1844 and caused a sensation.

The title of the picture already expresses its ambivalence. The Great Western Railway was the creation of the legendary industrialist and inventor Isambard Kingdom Brunel, the embodiment of all Victorian ambitions: engineer, man of action, entrepreneur, capitalist and genius. Turner's homage to the great man is double-edged, for the locomotive racing towards the observer, the only piece of solid material emerging from a cloud of rainy squalls and light reflections in a summer storm, seems lost amid this natural spectacle. The centre of the picture is empty. But the existential struggle is taking place between the locomotive and the now almost invisible hare, which, according to his contemporaries' accounts, the painter portrayed as the true incarnation of speed. The rain, steam and speed that give the painting its name turn the locomotive into an alien element.

At the start of the age of the railway, locomotives bore great names from Greek mythology: Actaeon the hunter, the sun god Phaedon or Cyclops, the one-eyed giant – they were all fire-breathing monsters on rails. Orion, the legendary hare hunter who never reaches his prey in the night sky, had likewise lent his name to a locomotive. According to myth, he had three fathers: Neptune, Zeus and Apollo, the gods of water, air and fire, respectively – the elements that move a steam engine, which create rain in nature; when technologically manipulated, they produce steam and finally speed: relentless in its land appropriation, its striving for technological solutions and its artificially fuelled race against history, from which it ejects itself.

Now the steam engine's hunt for the hare stands for a confrontation with the depth of time, as well as the shocking power of the all-conquering present. With a literal light hand, Turner added further

15 William Turner, *Rain, Steam and Speed – The Great Western Railway*, 1844, oil on canvas, 91 × 121.8 cm. National Gallery, London

playful possible signifiers. To the left of the bridge one sees a small boat rowing away from the viewer, as well as a group of figures standing on the shore, rather like bathers or ideal figures of the classicist type in the landscapes of Claude Lorrain. This is also an engagement with the classical legacy, with the ideal landscape and what progress does to it.

And the farmer ploughing his fields at the right edge of the picture? Is this another game within a game? Was Turner familiar with Bruegel's *Landscape with the Fall of Icarus* with the farmer in the foreground, subjugating the soil with his plough and dominating his bridled horse, yet enslaved himself? Had he seen prints of the original, or one of the copies made afterwards? Perhaps he even saw the original during his visit to Brussels in 1839? Is this the reversal of conditions described by Bruegel, the moment when a technical sun god comes close to the fiery star on his triumphant path while the farmer, now insignificant, can only watch from afar? Or does this Icarus still have a long ascent ahead of him before he too plummets into the waves?

This is a landscape painting about landscape painting; both compositions show a strong diagonal, a similarly tumultuous and enigmatic sky, a large water surface with boats on it, the barest hint of a city on the horizon. What sets it apart from earlier works of the genre, however, is that the hero of this new mythology is still racing towards the sun, for he has tamed the fire inside him and is flying on rails of steel. Turner seems to be continuing this conversation with a colleague and equally alert mind across the centuries.

The world of ploughshares and rowing boats has long been replaced by the brutal majesty of the bridge and the frantic change on rails, but the underlying warning is ever present. The pioneering age of the railways saw a particularly large number of terrible train accidents, and the dream of godlike velocity came with fatal risks – not only for the hare. The painter, who had still grown up in the eighteenth century, showed that his intuition and artistic horizon rooted him in classical antiquity, but that he had recognized the challenge of technological modernity and the lethal flipside of every apparent triumph over nature.

Modern Times

Storming obliviously into the future, the modernity that Blake and Turner captured in visions created its own answers to these images of the subjugation of nature – disturbing, distressing answers.

In Europe, the experience of the First World War marked a turning point. Until then, humans had dominated nature and used mechanical means to subjugate it. In the mechanized battles at the Western Front, it became clear to an entire generation that mettle, muscle and manliness or concepts such as honour and courage, principles, education or conviction did not stand a chance against the armada of machines sent to kill them. Anarchists and monarchists, Jews and antisemites, academics and workers, communists and stockbrokers could all cower together in the same trench until the same grenade, fired from a distance of several kilometres, tore them all to pieces and left their remains to trickle away in the mud, long since turned to liquid, of the heavily bombed battlefield.

Artillery and tanks, fighter planes and poison gas, machine guns and barbed wire: this first fully industrialized war (at the Western Front, at least) made it brutally clear how little the crown of creation could still control its own technological creatures; the machines had long surpassed their builders. Since early Christianity and even more since the Enlightenment, the goal of a civilized life had been to subjugate, control and civilize nature, which also meant suppressing or domesticating everything that was 'natural' – that is, base and libidinal – about humans in order to turn all of life into culture and devote it to culture, the endeavour of human civilization.

The logic of this narrative is so obvious that even the *Epic of Gilgamesh* already asked how a hero who is two-thirds divine and one-third mortal can overcome his mortality. In other mythical tales too, and in the Bible, humans are half-animal and half-angels or -gods, trapped between the two and endangered by their own irrational impulses.

The development of increasingly strong machines that could imitate human skills raises the question anew, and more urgently, since the human being is becoming a regrettable hybrid creature. Despite fasting and chastisement, sublimation and repression, it can never entirely shake off and dominate its animal half, and thus its tendency towards unreasonable behaviour. Theologically and technologically, humans remain fallible. Machines do not have this animal half, however; they are wholly automatic angels, wholly devoted to their purpose, designed to function smoothly. In a mechanical conception of the world, they were the next logical step towards a complete domination of nature, one more perfect than mere humans could ever have achieved.

The experience of the war made it clear to large numbers of people in Europe and its colonies that machines could seize control over nature from humans and rise to become rulers themselves, completely rational and perfectly constructed, free of desire, greed and even hatred – willing instruments that could become deadly weapons in the hands of sinister powers and potentially surpass humans by far.

There was an entire wave of popular culture in the 1920s and 1930s that dealt almost obsessively with the old relationship between humans and machines, a competition that humans seemed to be losing, slowly but surely. In Fritz Lang's monumental *Metropolis* (1927), a classic today but a flop at the time, a robot woman lures the masses to their downfall. In his play *Rossum's Universal Robots* (1920), the Czech playwright Karel Čapek tells the story of a line of robots designed as slave workers that eliminates humanity to take over the planet, and Charlie Chaplin showed in his masterpiece *Modern Times* (1936), choreographed with incredible precision, how humans, once installed by God to subdue the Earth, were now hunted and hungry factory workers who were literally being swallowed by the great machine.

The relationship between humans and machines was already described a century ago as a struggle for the complete domination of nature. As a natural organism, *Homo sapiens* had an evolutionary advantage, but machines caught up through a rapid process of development and seemed to close the gap. Then humans would finally have proved themselves to be dispensable, and the compromise that is the human being would be technologically obsolete.

The invention of the transistor and the internet, as well as the triumph of the smartphone, abruptly advanced this process and posed the question of subjugation in a new light. They transport their users into a world, often for many hours a day, that is governed by algorithms based on commercial interests and structured according to psychological insights, where they find recognition, affirmation, gratification, distraction and entertainment.

The old fantasy of humans as cyborgs that develop new abilities through implanted modules has thus become a reality, even if that module is usually attached to the hand rather than the head. That will come too, but the fundamental step has already been taken, now that human contacts and social connections of all kinds take place via platforms that openly build on the manipulation of their users and not only transform access to the world into consumer decisions, but also weight it according to which aspects of reality, which behavioural patterns and which weak points of human psychology and cognition can best be exploited to translate those constant interactions into usable data or transactions. If humans are half-angel and half-animal, to use an old formulation, there can be no doubt that this dominant aspect of digital communication is directed almost exclusively and specifically at the manipulable animal, at emotions and impulses, insecurities and needs – the ideal basis for the monetization of *Homo sapiens*.

The self-subjugation of humans has taken an unexpected path. For Saint Augustine it was the subjugation of our sinful, libidinal nature, and thus an immensely potent way to control the faithful themselves. Because this libidinal nature is an unparalleled source of income, however, it is logical in a capitalist context that it will be exploited as a resource in the same systematic fashion demonstrated by the church with confessions, penitential prayers, indulgences and control of sexuality.

Subjugation is no longer a struggle between human drives and the eternal soul. It has become a voluntary act – not in Hegel's sense, however, where free individuals freely decide to relinquish parts of their freedom to form a community with laws, giving them a power that makes many freedoms possible in the first place (an aptly Hegelian bundle of paradoxes). The self-subjugation of comparatively wealthy people to the laws of digital interconnection is in effect the acquisition of a new organ, a powerful filter made of bright colours and commercially exploited options that places itself between individuals and their environment, a filter constructed in such a way that it constantly urges, invites and demands new interactions, then sweetens them with little rewards.

In the rational, rationalized world of modernity, there is no longer a God who commands humans to subdue the Earth. God has been eliminated, but not the human longing for him – or, at least, for a meaning beyond one's own trivial existence. Nietzsche's murderers of God, stumbling under the burden of their crime, and Freud's neurotic repressors of their own unacceptable desires, have seemingly come to terms with their defeat. The rule of machines completes this historical logic, which becomes even more dynamic and penetrating through Artificial Intelligence and will lead into a posthuman age. Humans are increasingly losing not only the means of domination to the necessities of digital interconnection, but also the possibility to understand the decision-making structures and processes of this digital world, meaning that they tend to be existentially at the mercy of the logic of things. This is not necessarily perceived as an existential problem, however. The sorcerer's apprentice in Goethe created chaos and could not dispel the spirits he had summoned, but he has also discovered Tinder, and currently has better things to do than grapple with crazed broomsticks.

Victory ends in defeat, and J. M. W. Turner already saw it coming while the world of steam locomotives and engineers was still celebrating its exuberant successes. Never had subjugation been so complete, so incontestable, and never had the advances been more epic. At the same time, these frantic pistons no longer had the bearings to keep them running smoothly. Every metaphysical reference point for human endeavours had become brittle and disputable, and fearful, guilt-laden patricides were helpless to fulfil the task of restoring an objective morality to the world. Progress was a brutal business, but the most eminent scientists and most popular authors of the time promised a heavenly Jerusalem, a time of total peace in which, thanks to science, hunger, poverty and war would be a thing of the past.

COSMOS

Agony

Never was the subjugation of nature celebrated more consistently than in the post-war years, between the nuclear bomb, the moon landing and the oil boom. The capitalist West and the communist states competed for the greatest, most spectacular successes in a proxy war of dams and shaft towers, factory landscapes and dead rivers, blast furnaces and unfathomably deep mines.

The environmental balance of the economic boom was alarming from the start, but has now been distorted into a system of organized, deadly madness in which a primate went forth to destroy its own natural resources. At the same time, the societies that propelled this destructive transformation became masterful at reshaping the faded religious imagery into an illusory new reality, breathing new life into religious images in commercial contexts to create a new, calming ritual practice.

The apologists for religion were followed by the apologists for the market, the heirs of the theologians and historians and professors who had long practised the artful conversion of manifest contradictions into beautiful pictures. As the growing economic needs and possibilities caused increasing destruction, the respective societies also became more efficient at presenting the consumers – that is, themselves – with a professional, highly polished and superficial reflection: the world as product line and projection.

But the real world behind the scenes continued to exist, and it was restless, in constant but often microscopic billionfold upheaval, a cascade of complex transformations that could not be adequately represented by science, were not considered relevant economic factors and were politically insufficient to draw anyone from the woodwork.

Modernity proceeded with the 'creative destruction' of existing connections so admired by Friedrich Hayek, whether social, historical or natural. Systematic campaigns for the subjugation of nature were already launched with impressive self-assurance in the mid twentieth century. 'The destiny of man is to possess the whole earth', wrote John Widtsoe, director of the US Federal Bureau of Reclamation, in 1928. He continued: 'and the destiny of the earth is to be subject to man. There can be no full conquest of the earth, and no real satisfaction to humanity, if large portions of the earth remain beyond his highest control.'[1]

The striving for absolute human domination was not unique to capitalism, however. The triumph of the proletariat over its bourgeois and capitalist oppressors was merely a historical prelude to the decisive battle against nature as the final frontier of human development. 'We cannot expect charity from nature. We must tear it from her', screamed a party slogan, and Leo Trotsky, temporarily Stalin's great rival, proclaimed a socialist future of total power:

> The present distribution of mountains and rivers, of fields, of meadows, of steppes, of forests, and of seashores, cannot be considered final. [...] Through the machine, man in socialist society will command nature in its entirety [...]. He will point out places for mountains and for passes. He will change the course of the rivers, and he will lay down rules for the oceans.[2]

Not to be outdone by so much utopian energy, Stalin imposed a literally murderous programme of construction projects on the Soviet Union that would not spare any force of nature or environmental structure. In his 'plan for the transformation of nature', proclaimed in 1948, the General Secretary of the Communist Party decreed the transformation of the Volga into a series of reservoirs. Six million hectares of land and countless lives of forced labourers disappeared beneath the artificially trapped tides. Two other rivers, the Ob and the Yenisei, would be diverted from their northward course to irrigate areas farther to the south.

As was already the case in the hydraulic societies of the Bronze Age, irrigation was the prerequisite for population and economic growth, and was now also connected to energy production. From the Aswan Dam in Egypt to the Hoover Dam in the USA or China's Three Gorges Dam, gigantic waterpower projects dominated landscapes and living conditions on all continents. 'Man must conquer nature', as General Secretary Jiang Zemin declared at the inauguration of the Three Gorges Dam.[3]

In the great communist dictatorships, people followed propagandistic ideas like religious laws, often out of fear of criticizing superiors. At the behest of the central bureau in the USSR, rivers that flowed into the Aral Sea were diverted to boost cotton production in Tajikistan, Turkmenistan and Uzbekistan. Industry was thirsty, and by 2012 the Aral Sea, once the fourth-largest freshwater lake in the world, had shrunk to a tenth of its original size.

The messianic dream of the dictatorship of the proletariat left a trail of natural destruction, contamination and dying landscapes in its wake. China experienced its greatest humanitarian disaster after the 'Great Leap Forwards', inspired by such declarations as 'Chairman Mao's thoughts are our guide to scoring victories in the struggle against nature' – a struggle that ended with a bitter lesson about the systemic reactions of natural cycles. To protect the fields from predatory birds, billions of animals were hunted and killed. The following year the insects came, especially the locusts, which barely had any natural enemies left and completely eradicated the crops; millions of farmers starved to death.[4]

The subjugation of nature has no ideological colour, no religious source, no political conviction. It was born in Asia, grew up in Europe and became a global citizen long ago. Colonialism carried it into the world, but for postcolonial governments and liberation movements too, its merciless rules stipulated that the maximum exploitation of natural resources was the only way to survive and measure success. Communists and capitalists, religions of every inclination, smugglers and investors, corrupt politicians and warlords and even democratically elected people with the best intentions contribute to the destruction. It is a function of an enormously increased population of *Homo sapiens* with even more increased needs and technological possibilities.

It was only after the devastation of Hiroshima and Nagasaki by nuclear bombs that doubts arose. The title of one of the most important books on the subject, Robert Jungk's *Brighter than a Thousand Suns* (1956), already evokes the *horror vacui* of the humans who irreversibly interfered with the internal order of nature and desecrated what is most holy. Taken together with the Holocaust as an evil apotheosis of industrialization and instrumental reason, as outlined by Theodor W. Adorno and Max Horkheimer in their *Dialectic of Enlightenment* (1947), both events became emblems of horror at the consequences of the absolute self-empowerment of modern humans.

This self-empowerment proceeded apace. Despite the concerns of small groups of experts and activists such as the Club of Rome, whose 1972 report *The Limits of Growth* presented a theoretical framework for public discussion that is still relevant, the top priority was boundless economic growth through maximum exploitation of nature. The decision to pursue prosperity and growth was all too understandable in a world that had been through two global wars and long periods of colonial occupation, and in which new nations were coming into being. This striving for power and prosperity was as old as history itself; what was completely new was the possibility of realizing these ambitions. This potential bubbled up from the ground.

Hard coal had fuelled the Industrial Revolution and led to a staggering increase in production, populations, wealth and power projections among the nations that allowed this revolution to explode. The raw material brought forth by miners from the innards of the Earth advanced the transformation of entire societies and created a new world, a new social order, new possibilities and new political dreams within a few generations. At the same time, an uncomfortable but relatively stable equilibrium developed between workers and their bosses. Coal mining was labour-intensive and resisted automation for a long time. Miners and other workers had to toil for long hours in terrible conditions, but they could strike for their rights, for if their strong arm stayed still, the wheels would stop turning.

Then a different natural resource reshuffled the pack. The extraction of crude oil did not require an army of miners; far fewer workers were needed, and it all took place far away from Western societies and their quarrels over human rights and conservation. Crude oil was cheap, easy to transport and endlessly versatile as the basis for products from chewing gum to kerosene, and it concentrated production and marketing among fewer and fewer companies. Crude oil drove the globalized world and simultaneously turned it on its head. Factories no longer had to be built in places where coal could be found, since oil could reach anywhere. Heavy oil fuelled enormous container vessels so cheaply that raw materials and production stages could be spread between any number of locations depending on working prices, labour laws, conservation, tax regimes or the venality of different contenders. The progressive subjugation and exploitation of nature escalated, but this dirty dimension of progress was suppressed by the countries that profited from it but were thousands of kilometres away.

The world as projection and product line: heating oil and diesel, petrol and kerosene, PVC and nylon, natural gas and coal allowed a handful of generations to emit the fossil energy stored over millions of years in the bodies of innumerable organisms and conjure a festival banquet of consumerism from the process. At the same time, these innovations caused a shift of global power relations – not only with regard to the oil wealth of the Middle East and the Russian natural gas reserves, but also through the tempo of capital, which moves faster than humans ever could. The regime of global production left the workers in regional economies defenceless. Capital could be transferred in the blink of an eye, factories quickly built elsewhere and production relocated; there was always someone cheaper, and the same people who lost their jobs in the factory were forced to make consumer choices that accelerated their own economic disempowerment and marginalization. No one needed expensive European coal any more, and production was cheaper elsewhere.

Pragmatically speaking, it was sensible to submit to the centrifugal force of the economic boom and technological miracles of the twentieth century. It would have been almost unimaginable, and probably career suicide for anyone in politics, to reject technological progress amid this euphoric sense of new beginnings. Thus, in the rapture of a post-war era that had just left behind unknown horrors and celebrated historic triumphs, a new, final vision of the domination of nature was initiated with incredible enthusiasm: the eternal hegemony of the markets, the proletariat, freedom – and the saviour.

In the momentum of this dynamic, certain theological assumptions were taken up into economic theory. In many Christian traditions, humans are not really part of nature. They are installed by God to emancipate their immortal soul from its bodily shackles, and their actions are carried out against the restrictions of nature and independently of it. In so doing, humans can choose freely between virtue and vice and act in the light of divine reason; they are ultimately instruments of a divine plan – namely, salvation from the earthly vale of tears.

The neoclassical economy that originated from Chicago and was especially dominant in Western states in the second half of the twentieth century likewise assumed that most economic activities take place independently of natural systems, that humans are rational individuals in possession of complete and relevant information and that they make decisions to maximize benefits. There is no psychological, anthropological or sociological school or study from the last hundred years that supports this extraordinary assumption in any way. On the contrary: there would not be an advertising industry if people decided freely and rationally.

Already in the early twentieth century, this irrational dimension of human activity became an integral part of new disciplines such as anthropology, sociology and psychology, which all knew through their own research that humans can only be understood as part of communities with specific cultural and social traits, that they act neither completely individually nor completely rationally, that they are hardly ever in possession of relevant and complete information, that they are affected by personal and historical traumas and can be influenced and manipulated. And not only that – as the war in Ukraine demonstrates, even people with a great deal of money and power occasionally take actions with far-reaching consequences out of motives such as guilt, humiliation, revenge or vanity – factors that should not play a part in a rational market.

The most important economic school in the West clung to a conception of humans that seems to have been taken directly from the theological tracts of the Middle Ages or early Modern Age, and ignored the scientific insights of its own time. Like Descartes's assumption that animals have no soul and no feelings, the conception of humans in this school of thought flies in the face of all evidence and experience, but it has the tremendous advantage of reducing human behaviour to such a small set of variables that it can easily be represented in mathematical models. We should bear in mind that this model has barely any connection to reality and empirically examined patterns of human behaviour, but at least one can calculate projections. 'That may work in practice', an old joke among economists goes, 'but does it work in theory too?'

The human being in neoclassical economics is a highly theological concept. But the world in which it acts also proves to be a construction from theology – namely, a dead territory that humans can systematically and profitably exploit and use as an unlimited economic resource, without fear of consequences or feedback effects. For much of the twentieth century, environmental factors such as clean water or clean air were treated in theories simply as external, as factors that did not need to be taken into account in modelling economic cycles and planning business models or political economies, since they are essentially universally available. Humans act outside and above nature.

After all, the historical thinking of this school, as found in the works of Francis Fukuyama and others, has a decidedly messianic or teleological character that sees the course of history as oriented towards a goal – one that, quite by chance, coincides with the truth and virtue of the school's own principles, since it consists in organizing all societies as liberal democracies in liberal markets, in accordance with the inexorable pull of history. That is the theology of globalization.

The path to this new Jerusalem was taken in leaps and bounds from around 1960. From that point on, with the aid of mathematics and machines, humans manipulated natural systems with an efficiency and speed that surpassed anything seen before: the extraction of fossil fuels, the emission of CO_2, the global population, the meat industry, science and innovation, plastic waste, global temperatures, arms production and drug trafficking, private fortunes and major corporations, antibiotics and fast fashion, tourism and energy consumption grew explosively until the pandemic, and there is no indication that things will be different in the future.

But there are more aspects of industrial and post-industrial societies that would be unthinkable without the immense economic and technological surge of the twentieth century. The successful struggle for equal opportunities for women and minorities, 1968, the sexual revolution and the strengthening of individual versus collective rights would scarcely have been possible without the oil boom, since the liberalization of moral principles, the loosening of patriarchal structures and even the tolerance for differences developed with growing prosperity and social stability. One has to be able to afford human rights, it would seem, especially since they are worthless unless they are enforceable and institutionally secured.

This is the crucial dilemma represented in a famous graph, known as the 'hockey stick', which shows how global temperatures developed in the previous century, namely in the form of a hockey stick propped up against a wall, with a curve that starts gently and becomes increasingly steep, finally shooting skywards. When the climatologist Michael Mann made his calculations available for publication by the IPCC (Intergovernmental Panel on Climate Change) in 2001, he was harshly attacked by colleagues. He was accused of using outdated data based on statistically dubious procedures and employing his unscientific results for propagandistic purposes. It soon became clear, however, that many of his harshest critics worked at institutions financed directly or indirectly by the fossil fuel industry. Independent investigations later confirmed Mann's results several times over.

A graph published by the NASA space agency reproduces Mann's hockey stick, albeit in a considerably larger context (figure 16). Analyses of ice cores were used to reconstruct the level of atmospheric CO_2 over the last 800,000 years – four times as long as the existence of *Homo sapiens*. The result is a constant fluctuation between ice ages and periods of warmth – until 1950. That point marked the beginning of a different game.

The final, dramatic rise in the level of atmospheric CO_2 coincides precisely with the rising temperatures, as well as other indicators such

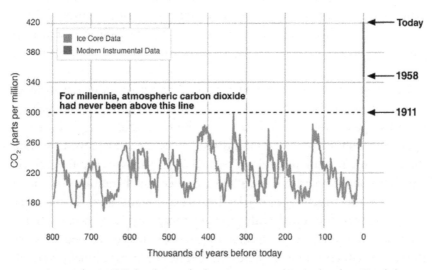

16 Atmospheric CO_2 level over the last 800,000 years until today. Graph by NASA. Source: https://climate.nasa.gov/vital-signs/carbon-dioxide

as the global population, economic output, meat production, waste, etc. This is the curve of human success, of successful subjugation. It points towards destruction.

The processual dynamics of a global economy whose prosperity and stability depend on constant and endless growth can be expressed in a few numbers. Before the pandemic, an area of rainforest the size of thirty soccer pitches was cut down every minute (the destruction has intensified since then), and currently a million tonnes of ice are melting in Greenland within the same time span, while the need for fossil-fuelled energy is still increasing in addition to the rapid developments in sustainable energy production. This is the economy of madness.

Meat production is a further indicator. Since 1960, the global population has more than doubled and meat consumption has grown fivefold; 88 billion land animals are slaughtered every year for the basic human right to a daily schnitzel, and those animals consume not only the soy from the destroyed rainforests, but also three-quarters of agricultural produce in Europe, as well as three-quarters of all antibiotics. Thus, factory farms have become an ideal breeding ground not only for breakfast eggs, but also for multidrug-resistant germs.

The One-Armed Lumberjack

This sudden, explosive growth within two generations left little time for reflection, but it is becoming clear that the answers of the post-war period cannot be answers today. Economic growth was the traditional post-war answer to every conceivable challenge. However, an economic growth of 3 per cent means that the respective economic output doubles every twenty-four years. Within slightly less than a generation, this means twice the demand for resources, twice as many products, twice as much pollution – or, with a miracle of sustainable technologies, a smaller but still catastrophic rise in levels that are already dizzyingly high. Economic growth can no longer be an answer to the catastrophe. This necessity of growth cannot be reconciled with the realities of an accelerating climate catastrophe.

This brings a key concept into play. It is difficult to follow the acceleration of history and understand its psychological meaning for the primate *Homo sapiens*. The first urban cultures in Mesopotamia came into existence 300 generations ago; 100 generations separate us from a walk in the streets of ancient Rome, and 10 from the French Revolution. But the immense transformations in the twentieth century – electrification and mass culture, Auschwitz and the nuclear bomb, urbanization and the oil boom, transistors and tourism, the internet and quantum mechanics, Artificial Intelligence and black holes – emerged within the span of a single stressed, unsettled human life.

There is a geological name for this period of rapid change: the Anthropocene, the terrestrial age of humans. The validity, time of origin and definition of the term are disputed, but perhaps it is primarily a matter for geologists whether it originated with the Trinity Test of 1945, when the first nuclear bomb was ignited and the first radioactive traces were left by humans in the planet's geology, or whether one should choose a different date. The fact is that *Homo sapiens*, in its laughably short history, has left considerable traces as a civilized being: a geological proof of progress.

In the light of this immense acceleration, it is all the more under-standable that at least one element in this age of subjugation, which has constantly been confronted with the unprecedented, offers a modicum of stability. As in the Enlightenment, this role was given to theological thought in a secular guise.

It is one of history's bitter ironies that, via colonialism, capitalism, socialism and anticolonial liberation movements, core ideas of an aggres-sively missionary Christendom became something like the operating system of a global world that prided itself on leaving behind religious and traditional ideas through scientific knowledge; the theological legacy is unmistakable. 'Humanity' stands outside and above 'nature', which it can and should subjugate and exploit for the realization of its goals, since this is its mission on Earth, its contribution to universal progress, the completion of history and the realization of its own potential.

The recent identifiable heirs to this messianic-theological tradition are the prophets of Silicon Valley, who seek their own culmination of history in various digital parallel universes or on other planets. The utopias of transhumanism likewise suppose that humans ultimately stand outside and above nature, and that it can and should be their goal to emancipate themselves more and more from natural connections. If an individual can be uploaded to the cloud based on purportedly relevant and trans-ferable data, then emancipation would seem to be complete: the perfect retreat to an optimized simulacrum of a mortal life's experience, the final flight of humans into the machine. Finally free of their ageing, ailing, lustful, reflexive bodies, they can live and experience themselves as pure spirit. Saint Augustine would have gone into ecstasy if he had had such means at his disposal, and Origen could have forgone castration if this digital salvation had sated his burning desire for disembodiment.

If one has been brought up to subjugate, it seems superhumanly difficult to perceive the world in any other way than through the all-distorting lens of a now defamiliarized, barely recognizable theological legacy – and thus one reads the world hierarchically, in the structure of master and slave. If reality no longer follows this schema, the conceptual rigidity produces something like the nameless frustration of the German Emperor Wilhelm II, who chopped down and sawed up trees like a madman in Dutch exile: over 40,000 of them, as revenge for his lost throne – and all with his right hand, as his left arm had been withered from birth. Unable to view the world and his role in it in any other way, he became Wilhelm the lumberjack. The former ruler of the German community of fate and nation of forest-lovers donated his neatly stacked firewood to the needy.

A story fails; a narrative drowns in the agony of consequences it can no longer predict or change. The great narrative of the subjugation of nature is thwarted by its side effects. This observation is not an invitation to revel in Western, post-Christian, progressive guilt; the inhabitants of the rich northern hemisphere are, for the most part, no more stupid in their approach to the world than the last uncontacted tribes of the Amazon. Rather, the problem is that, for all their accumulated knowledge and technological capacity, they are not as intelligent or even wise as their cousins who think in different categories.

It is a lazy, sentimental cliché that 'natural peoples' are in possession of profound wisdom that arrogant, modern Westerners have lost. Undoubtedly, post-industrial societies can and must learn a great deal from indigenous knowledge and indigenous worldviews, and should approach it with all due curiosity and humility; but 'natural peoples' also exterminated animals and wasted resources, drained soil and destroyed forests, waged wars and maintained economies of suffering and destruction that – from a modern perspective – are difficult or impossible to understand. They too have been destructive, short-sighted and occasionally greedy, but their technological reach was so small that the consequences remained locally restricted and manageable, and their direct dependence on natural connections was so great that they could only survive through detailed knowledge of them. Their life and death were tied to a specific place and landscape, and the knowledge about those was crucial for survival.

Globalization has turned this situation on its head. Actions are still individual and local, but their collective effects take on global, systemic and even interplanetary proportions. Economic relations have long since become dependent on global production chains and capital flows that often sever all connections to a specific place.

There are no more local problems or local actions. Every T-shirt connects people on different continents and different places in the production chain with land appropriation and the destruction of the rainforest; the power of agricultural giants and the effects of fertilizers, pesticides and other chemicals on rivers and groundwater; the collapse of insect populations; global transport networks and sweatshops; endlessly proliferating slums; children and teenagers ruining their health on production lines in dangerous factories; with offices with panoramic views in Shanghai, Houston or Lagos and Swiss bank accounts of useful partners in authoritarian governments; with container vessels fuelled by heavy oil, shopping malls with enormous carparks and their own artificial climate (at least one stable thing, as long as they have electricity); and with the endless rubbish dumps on which almost 80 per cent of fast fashion lands, and where children fight over meals with wild dogs and seagulls. Every scrap of cloth is part of dozens of global networks of profit and hope, of exploitation and deforestation. Every item of clothing bears the imprints of dozens of hands.

This interconnected point of view leads to further insecurity in a world already bursting at the seams, for it is impossible to view all everyday decisions as world events connected to unknown realities by 10,000 subtle webs. The world as product line and projection offers relief from the paralyzing, auto-aggressive feeling of infinite entanglement. A look in the mirror is enough for us to be overwhelmed by our own inadequacies in a self-enclosed and eternally competitive system of consumption, self-optimization, social media and work to such an extent that any thought of the world behind the mirror is deferred indefinitely.

Nietzsche wrote that this economic hall of mirrors is merely 'a surface- and sign-world, a world turned into generalities and thereby debased to its lowest common denominator'. The signs always float on the top, always stand in the foreground and are legible as the first or only elements, such that 'everything which enters consciousness thereby *becomes* shallow, thin, relatively stupid, general, a sign, a herd-mark'.[5] In the USA, with its mythical cowboys, one called such herd-marks brands, using animals to describe the mechanism that makes the present legible in many places.

Perhaps this legibility through branding and consumer choices is so strong because – in addition to obvious motivations such as greed, vanity and herd instinct – it is built on very old rituals. The female 'consumer' – surrounded from morning to evening by images and messages suggesting her inadequacy and the existence of a better world full of ideal, happy, young, rich, content and attractive people – is not so far away from the farmer's wife 500 years ago who received variants of such messages from the pictures of saints in the church, which likewise spoke of a better world and a higher form of life.

However, the farmer's wife and her distant granddaughter in the city are connected most of all by a similar path to salvation, to a temporary rescue through sacrifice and participation, through prayer and communion. What the church was to the farmer's wife is now represented by the shopping street, where the promise of participation in the ideal world is fulfilled in an act of communion consisting in the exchange of money and branded commodities. At the end of this deeply religious process is one's own identity, consisting of legible gestures, labelled with trademarks and supported by marketing campaigns, which define an individual in the eyes of others.

Of course, unlike the farmer's wife, her granddaughter needs money for her participation in the transcendent world of happiness. The identity she thus acquires remains eternally unstable; it must always be re-filled and re-validated through new acts of communion, new purchases of legible products with a degree of built-in semiotic obsolescence. There comes a point when toasters and car parts are broken and can no longer be repaired, only replaced by new products, and the signal effect of products has an even shorter shelf life. You might as well be dead as out of style, as the saying goes.

It is hardly surprising that societies which are so dominated by religious practice and old, theologically created notions find it extremely difficult to liberate their thinking sufficiently to recognize the outlines of their own, systematically nurtured insecurities behind the subjugation or taming of nature that has been elevated to the goal of society.

Liberal Lifelong Lies

The psychological burden of re-assessing one's experience is further increased, especially in Western countries. Until the collapse of the Soviet Union, it seemed that the world was bipolar and any future was thus a clear choice between liberalism and (Soviet) totalitarianism. This was not a difficult choice for the majority of people in the West (and many in the East), since the living conditions and expectations, the respective freedom or unfreedom, were obviously very different on the two sides of the Iron Curtain, and an embrace of the Soviet model required sufficient intensity of ideological conviction to take on board the death of facts and enemies, and not infrequently friends too.

Then came the fall of the Iron Curtain, and with it the self-assurance of triumphant liberal messianism. After 1989 it seemed that, to quote Francis Fukuyama's famous formulation, once again the 'end of history' had been reached, and liberal observers were fairly sure that the whole world would now, carried by the invisible current of world history, inevitably be washed into the safe haven of democracies and free markets, irresistibly drawn by the superiority of liberal markets and the liberal politics that had to follow them.

The opposite happened. Liberalism only continued to prevail in the form of new variants such as kleptocracy and illiberal democracy, while many societies turned their backs on its democratic expression, and neither all the money nor all the power in the world could convince Iraq or Afghanistan to choose the liberal path. The Western model had a limited attraction as an export, it seemed, and had long been in crisis in Western countries too as a result of deindustrialization and globalization.

Two further factors undermined the liberal gospel. The first is economic output, the much-lauded core competency of liberal societies, which faltered considerably in the crisis of 2008 while authoritarian countries like China took millions of people out of poverty without democracy or free markets, contrary to all Western wisdom, and showed that perhaps democracy cannot survive without markets, but markets can certainly survive without democracy – and met with relatively broad agreement. Although one should take opinion polls on political questions in countries such as China with a pinch of salt, observers have stated time and again that the majority of the Chinese population are genuinely content with the country's governance and reject accusations concerning freedom of expression, capital punishment and oppression of minorities; instead, they point out low crime rates, newfound prosperity, reduced corruption and growing national pride. Clearly, liberal democracy is no longer the form of society that is automatically the best because it functions best.

The second undermining factor is the lifelong lie of the liberal project itself. Anyone who grew up in a Western country in the post-war period learnt from school, television and newspapers that the immense and visible progress of Western societies – their prosperity, peace and relative social equity – were the result of hard work and punctuality, democracy and transparency, enlightenment and education, community spirit and separation of powers (which is precisely why the Holocaust was so difficult to integrate into this picture and ultimately had to be sacralized so that it would no longer be an obstacle to the normal functioning of progressive morality).

The other side of this progress narrative – from the significance of slavery and the lumpenproletariat for the Industrial Revolution to dirty proxy wars, from America's genocidal land appropriation to colonial mass murder and the poisoning and destruction of entire societies and swathes of land – did not feature in history. That is one reason why liberalism seemed like such an illuminating response to the dark machinations of the communist enemies. However, even a small historical distance is enough to see how the liberal project has been compromised by its political, economic and intellectual elites and by environmental destruction through mass consumption.

In the fluidity of the present, reference points that were stable for generations are no longer stable or have disappeared entirely beneath the surface, and the former historic mission of humanity – another narcissistic valorization – has been negated by reality. It feels good to view one's prosperity as the just reward for one's own hard work. As the debates of recent years have shown, however, one can even gain moral gratification from the opposite: it can also feel good to be a member of a privileged minority and accuse oneself of every possible historical and economic guilt, and thus emerge from this discursive catharsis morally cleansed and in possession of the ultimate truth. The collective mea culpa of liberal elites induces some to confuse symbolic restitution with political or economic action.

Faced with such massive uncertainties, however, it is understandable that the public discussion clings to mindsets and ideas adopted somewhat unthinkingly from the distant past, whose allegorical understanding of the world is no longer adequate to the challenges of the Anthropocene. Although they are obviously false as far as the scientific view of humans is concerned, they nonetheless have the huge advantage of being something familiar in a chaotically developing world, belonging to a sort of cultural bloodstream and painting an extremely flattering picture of the significance of humans vis-à-vis the rest of nature.

This flattering picture conceals the fundamental paradox of historical success in Western, and subsequently global, civilization: immense technological and scientific progress became possible not only through moral compromises at the expense of others, but also because the physical and biological destruction it caused were ignored and it effectively destroyed the basis for its own existence. This led to a sort of salvation history of the market, which stipulated that the entire historical development was striving towards its own ideal resolution – and whose argumentative substantiation financed a growing industry of justification, entertainment and projection that applied theological thinking with a new vocabulary and old images to the present day.

The concept of agony seems appropriate for that dramatic moment in the hero's life when his greatest lifelong dream is thwarted and he reaches the lowest point of his path, like Christ's agony in the garden of Gethsemane and then on the cross. But is agony not followed by salvation?

The World as Clockwork

But as I finally slowly, slowly paint this gloomy question mark [...] it strikes me that I hear all around myself most malicious, cheerful, hobgoblin-like laughter: the spirits of my book are themselves descending upon me, pulling my ears and calling me to order. 'We can't stand it anymore', they shout, 'stop, stop this raven-black music! Are we not surrounded by bright mid-morning? And by soft ground and green grass, the kingdom of the dance? Was there ever a better hour for gaiety?'

Friedrich Nietzsche, *The Gay Science*[6]

No: no salvation, no heavenly Jerusalem. Instead, a disaster and a revolution greater than the Copernican one; a humanity-defining event, here and now. Instead, the end of 3,000 years of cultural history and the beginning of a journey into unknown universes – as a being almost unknown to itself.

But not so fast: those are big words. They should be treated as carefully as a grenade with the pin removed. So what happens if our rehearsed understanding of history no longer gives us meaningful access to the world – if an entire worldview goes blind? Stories can be shattered by reality, and the stories conquerors and autocrats tell themselves were evidently not always entirely counterproductive, otherwise they would not have got where they did. But it is possible, even common, for a story to lose contact over time with the reality from which it arose, especially if circumstances change. Sometimes the story itself even changes the circumstances, as with the West, whose doctrine of subjugation was expedient until the use of oil caused its successes to explode exponentially more rapidly than the possibilities for reflection.

The period of critical and catastrophic developments, which has already begun and will determine the foreseeable future, cannot be survived based on the structures of subjugation, domination and growth, since this perspective no longer has any useful answers to offer. At the same time, this history has been so successful for centuries, and has displayed such a strong transformative power, that it seems almost impossible to discard entirely the conception of the world based on subjugation and the conception of humans based on a usurper standing above nature.

Let us stay close to this problem area in the biography of a contagious idea at the moment of its death and the inescapable question of what comes next. Which idea, which worldview, which grand narrative has the potential to bring humanity out of an existential crisis, a crisis that already manifests itself in a dramatic collapse of natural systems? What can the moment of this narrative's death reveal about the future?

Let us think this revolution in images and begin with an immense, complex automaton of steel, gold and brass, with countless delicate cogs and shining blue springs, with levers and shafts and pointers and buttons on a surface where figures, connected to the inside of the machine by hidden cogs, move about and reel off the everyday drama of societal life. Let us think this image on every scale, from the functioning of the genome to the construction of individual cells and finally the mechanics of entire galaxies.

This image of the world as clockwork, which we encountered in the first chapters, runs through the Enlightenment and modernity and became immensely potent. But this metaphor had a history too, and only established itself slowly between the discovery by Nicolaus Copernicus (1473–1543) that the Earth must revolve around the sun and the late eighteenth century, when a scientific method had been sufficiently recognized for a self-enclosed worldview to arise from it.

The key to this new understanding was a shift of perspective. Well into the Renaissance, the world was viewed by scholars as God's creation, which made theology an unrivalled instrument for understanding it. Empirical findings or mathematical calculations could be helpful, but divine revelation and the biblical account of the creation remained the ultimate authority on any matter. If fossil finds contradicted the creation story, for example, then they had to be subjected to sufficient interpretation to fit into the biblical order. In its inner workings, the world was as mysterious as the thoughts of God: an order that humans could not and should not understand.

Humans and their moral actions stood at the centre of this theological world, for it was up to them to fulfil God's will and thus the plan of history – as rulers and subjugators of nature whose different fields were grouped around humanity in concentric circles, as it were. Thus, the moral actions of individuals became the theological centre of the world in Christian traditions.

With the immense wave of development that swept through European societies in the seventeenth century, a new metaphor established itself. René Descartes's attempt to deduce the entire world, including a good, compassionate and almighty god, purely from reason caused him to see the world as a gigantic machine whose mechanism is endlessly complex, but essentially opaque. To use the machine for one's own ends, then, it was sufficient to reconstruct the lost instruction manual.

The undeniable charm of this mechanical metaphor was that it provided results and could describe causalities. Newton's laws described an observable reality in which everything functions according to rules. The second great advantage of this image leads to the second part of Newton's career, in which one of the greatest mathematical geniuses in human history immersed himself in biblical numerology, searched for hidden messages from the creator and tried to determine the date of the apocalypse. Religion and natural science were not as strictly separated then as they are today – though a surprising number of theoretical physicists still have a foible for mystical ideas.

Newton could reconcile his scientific laws with his mystical investigations because science was still describing a created universe whose creator had merely hidden his messages in the material world. This mechanical universe always left open the possibility of someone standing behind the mechanism who had conceived it, Voltaire's 'great watchmaker', if not the all-bountiful creator whom Leibniz sought to identify as a mathematician.

This idea of nature as a mechanism or a machine was perfectly suited to showing why objects fall down, why a parrot suffocates in a vacuum and why a frog's legs twitch when they are connected to an electric current. Mystically oriented truth-seekers criticized it as a simplification for ignoring higher levels of knowledge and reality, but its efficiency and explanatory power proved irresistible.

For scientifically inclined minds in a world that was still underpinned and pervaded by religious views, the mechanistic worldview had a further inestimable advantage. Immanuel Kant had argued that one could not draw any conclusions from the world of apperception about the mysterious sphere of 'things in themselves', and that science and religion could not come into competition with each other. The theologians took care of what lay behind phenomena, while the scientists could calmly explore the knowable world. A truce had been declared.

The universe as clockwork could also be wonderfully reproduced, and since the advent of mechanical clocks, mechanical models of the world have also been produced, often displaying stupendous complexity and artistry. These masterpieces of mathematical planning and supreme craftsmanship featured planets humming around the sun on fixed orbits. But if the planets could whirr through the air driven by hidden forces, it seemed likely that, as many Enlightenment thinkers claimed, what lay behind the mystery of life was merely a highly complex machine. Automaton builders also realized this idea in machines that were displayed all over Europe, such as a mechanical Turk that could play chess, that causal art of thinking in structures and decision trees. Hidden in this first chess automaton, however, was a short-statured chess master – the machine builders were not yet good enough after all. But their art was good enough for a multitude of figures and ships, all the way to mechanical birds in cages that could twitter thanks to hidden bellows. A mechanical duck could even eat bird food and excrete it in a seemingly digested form.

The world as clockwork is easy to classify and visually attractive. Impressed by industrialization, the German doctor Fritz Kahn took the idea a step further and, on a poster made in 1926, portrayed the human being as an 'industrial palace' with grey-haired planners wearing horn-rimmed glasses and white coats in its head, and a variety of devices, workers, wires and production steps in the rest of its body. This human could be seen as the heir to Descartes's soulless mechanical dog: a functional unit that can be effectively run, and repaired if necessary.

The mechanical conception of the world had another property that proved infinitely important. Humans had always tried to describe the world around them and force it into concepts, but the scientific method made hierarchies and family trees immeasurably simpler and more methodologically robust. Carl von Linné was able to classify all organic life (with humans at the top, of course) and other scientists dealt with chemical elements and geological epochs, social classes and ancient civilizations, 'races' of humans or species of jellyfish, greenback frogs or periods of art history. Everything was neatly divided and represented in ascending orders that almost always culminated in people's own bourgeois present.

Only a world that was classified and understood as a hierarchy and mechanism could be considered fully explored and conquered – and that is how it was shown by the conquering societies. Museums presented the hierarchy and classification of a universe in which every little cog has its place in order to fulfil the historical mission of domination. In the second half of the nineteenth century, this dominant understanding showed itself in its purest form in Vienna when the ring road was built. The Art History Museum and Natural History Museum are placed, around the square named after Empress Maria Theresia, at an appropriate distance from her statue: culture and nature perfectly classified, framed allegorically by classical nudes and great minds and divided into orders and genera, periods and materials, floors and rooms.

There we encounter old acquaintances from theological debates of bygone centuries: humans elevated above nature as the crown of creation; civilization as emancipation from natural needs; the rational structure of the universe; the universality of one's own truth; the teleological view of history moving towards a goal; the great project of completing domination. This is also part of the mechanical conception of the world: every machine has a function, a purpose to fulfil. For how could the universe exist without a constructor or a goal?

Images guide or force our thoughts in particular directions and suggest particular conclusions. As soon as one thinks of the world as a machine, the logic of the mechanical is inescapable. Every part of the machine, every simple spare part, is a smallest unit; it cannot be broken down any further without losing its identity. A cog or a screw is simply a cog or a screw, and a part of it is merely a functionless fragment. It consists of a material and can be melted or otherwise divided into its component parts, but it has its smallest functional life as a cog. Everything one needs to know about the cog as a cog comes from its form and from its place and function in the machine. Anyone who makes the effort to analyse every cog, shaft and spring individually will ultimately understand the entire machine.

This worldview produced fine models and pictures, but could only explain one part of the observable phenomena that had occupied philosophers for a long time. If the world only consisted of atoms, then where and what was the soul? How could life come into being? What filled the cosmos? How could humans be sure of their truth, and what relation could there be between God and a purely material world? A number of intellectual support structures were developed to plug these holes in the projected truth, from the fluid and ether of the eighteenth century to Hegel's world spirit and Kant's things in themselves.

Admiration for Cannibals

The planet as divine creation and firmament, the universe as clockwork – these images are familiar to us, but they construct horizons of understanding that cannot inspire any productive action in the present crisis.

Understandably, the antithesis of these models is less familiar, since Western humanity in particular successfully inhabited and applied these models for generations, because every person who grew up in a Western society has to some extent learnt and internalized that the world should be understood from this perspective and using this grid. Nonetheless, attempts to grasp the world from a different angle were always part of reflection on nature, even in Europe and the so-called West, where subjugation had conquered most people's hearts and minds.

While a dominant opinion within Western societies had completed the intellectual separation of culture and nature and built its self-identity on it, there were always some attempts to escape this normative separation with its trail of philosophical and political consequences. As early as the late sixteenth century, an observer such as Michel de Montaigne did not even try to buttress the gaps in his knowledge of the world using philosophical scaffolds; on the contrary, he focused his interest on exploring them.

Why, he asked in his essay 'On the Cannibals', was the opinion of Europeans about indigenous South Americans who visited France (and were actually brought back again by their hosts) more interesting than the observations of indigenous peoples about French society? Considering the bloody religious wars waged in Europe in the name of a merciful God, who were the real barbarians? And when his cat played with him, was he passing the time with it or vice versa?

Montaigne was prepared to let his scepticism towards all accepted truths lead him to ask further questions, starting with the exclusive divine gift of reason. What if this reason on which he had built 'all these great advantages that he thinks he has over other creatures' was not actually so astonishing or so unique? Was he not letting flattering fairy tales go to his head?

> Who has persuaded [man] that that admirable motion of the celestial vault, the eternal light of those torches rolling so proudly above his head, the fearful movements of that infinite sea, were established and have lasted so many centuries for his convenience and his service? Is it possible to imagine anything so ridiculous as that this miserable and puny creature, who is not even master of himself, exposed to the attacks of all things, should call himself master and emperor of the universe, the least part of which it is not in his power to know, much less to command?[7]

How can one trust a reason that is convinced by such a transparent game, and 'this privilege that he attributes to himself of being the only one in this great edifice who has the capacity to recognize its beauty and its parts, the only one who can give thanks for it to the architect and keep an account of the receipts and expenses of the world: who has sealed him this privilege?'[8]

This instrumental worldview is quite simply ridiculous, Montaigne claims. The majesty of the universe and its celestial bodies cannot be compared to humans, who spend but an instant on this planet and flatter themselves that they control it, despite being driven by influences they themselves cannot grasp. What they are missing in particular, however, is the obvious and theologically underpinned narcissism of only believing humans capable of a complex existence. Considering the stars, he asks himself:

> Why do we deny them soul, and life, and reason? Have we recognized in them some inert, insensible stupidity, we who have no dealings with them except obedience? Shall we say that we have seen in no other creature than man the exercise of a rational soul? Well, have we seen anything like the sun? Does it fail to exist, because we have seen nothing like it, and its movements to exist, because there are none like them?[9]

Montaigne did not quite open the gates to hell with such remarks, but certainly the door to parallel universes. Since animals react in the same way as humans to the same impulses and insist on having their way, he argued, we must confess that 'this same reason, this same method that we have for working, the animals have it also, or some better one'.[10]

Humans are incorrigible dreamers, Montaigne writes: ever succumbing to the temptation to relate everything to themselves without stepping back from their own perspective, then presenting themselves in an unrealistically good light. If humans and animals are not as different as humans like to flatter themselves that they are; if even the universe can have its own intelligence, a form of life that is neither provable nor refutable; if humans react to the influence of planets and unseen impulses and far more actors are connected by intelligence and consciousness than is allowed by the theological worldview – what does that mean for the place of humans in this universe?

Montaigne was intellectually daring, but not suicidal. He left such questions not only unanswered but also unasked, although they are obviously the next step after his arguments, which are decorated with venerable classical quotations. Nonetheless, he was not the only one whose thinking tended towards such radical conclusions. Bernardino Telesio had attempted in the sixteenth century to describe the world as a great organism, but his image of a living planet had not won out over the metaphor of nature as a machine. Spinoza had argued for the connectedness of all being based on an infinitely varied communicating substance, but was banished from the canon for being too dangerous. The world of the Enlightenment and the Industrial Revolution remained an automaton, a factory, a machine.

Here one can recognize the first outlines of a worldview that thinks in networks rather than hierarchies; not in boundaries, but rather connections and spaces of communication; not in individual events, but rather the entanglement of complex symbioses.

The first substantial scientific attempt to describe the entire universe as a system of interconnected systems remained incomplete; perhaps there could be no productive completion. Alexander von Humboldt (1769–1859) was a truly universal mind and scholar who had spent several years travelling through Latin America, Central Asia and the USA for fieldwork, and his alert intelligence had found fascinations and connecting points everywhere. He researched and wrote about physics and geology, about plants and minerals, animals and climatic

248

conditions, astronomy and anthropology, chemistry and archaeology, making significant contributions to all of these disciplines.

Humboldt's travel reports revealed him as an observer who was genuinely prepared to engage with what he saw. In a time when readers no longer admired anything except classical antiquity, he described the sculptures of the Mayas as equal or even superior to those of the ancient Greeks, investigated the secrets of Aztec hieroglyphs, examined the details of the Aztec calendar and observed, between the populations of the different areas, 'differences as striking as those between the Arabs, Persians, and Slavs, all of whom belong to the Caucasian race'.[11]

In his observation of other cultures, Humboldt did everything that Hegel and other so-called great thinkers had never found necessary: he created an empirical basis for his opinions, which meant that they changed repeatedly and moved ever further away from the intellectual horizons of his contemporaries:

> Venturing back to the earliest times, historical research reveals to us that nearly every part of the earth was once occupied by men who believed themselves to be aboriginal because they were unaware of their filiations. Amidst a multitude of peoples who have succeeded one another and have intermixed, it is impossible to determine what exactly was the initial base of the population, that original layer beyond which begins the domain of cosmogonic tradition.[12]

These were fighting words in a Europe where much of the intellectual avant-garde thought in nationalist terms. But Humboldt had another message to his contemporaries and their striving for national and colonial greatness: 'Nothing is more difficult than comparing nations that have taken different paths in their social development.'[13]

The insatiably curious Humboldt hated nothing more than lazy generalizations. His descriptions of landscapes, artworks, plants, volcanoes, people, monuments and climate zones are always specific and detailed. They describe physical characteristics and historical aspects, point out different languages and customs, seeing or searching for the richness in every form of life and every cultural expression. His scientific method consisted not only in measuring precisely, collecting, evaluating, reading and publishing, but especially also in conversations with colleagues and friends on long walks, in over 30,000 letters and countless visits.

It was only logical that Humboldt chose a completely impossible task for his life's work: to describe the world as a whole, as a system of systems, united in mutual dependence and the breathtaking beauty of all life. He spent almost his entire life as a scholar working on his idea for a great work in which he could finally bring everything together: *Cosmos*. Almost as if he had wanted to follow on from Montaigne's long Latin quotations, but also with the classical tradition of the German bourgeoisie in mind, he took a quotation from the natural history of the Roman thinker Pliny as his motto: 'In truth the power and majesty of Nature at every turn lack credibility if one views these aspects piecemeal and does not embrace them as a whole.'[14]

This, Humboldt argued, was the problem with thinking in cogs and screws: it focused knowledge on individual parts and isolated competencies, describing these parts as interchangeable, purely functional and irreducible. Humboldt's travels and studies had caused a very different picture to develop in his mind. Instead of retelling this classical tale, Humboldt chose a different path. He wanted to show the different aspects of the planet in their interactions: the effects of the climate on plant species and human cultures in different areas and at different altitudes, the firmament and human knowledge acquisition, rock layers, volcanic eruptions and the first plants that return after them, the history of physics and the astronomical ideas of his colleague Laplace – everything aroused his interest and his will to test for similarities and peculiarities through comparisons.

Humboldt never completed his great *Cosmos*, but he had already said what he wanted to say and could say with the means of his time, and it was at least a considerable commercial success – a bestseller even. *Cosmos* was a disappointing read, however; the great scientist was less impressive in his writing style. He got lost in complicated sentences, endless tangents and reflections as well as enormously long footnotes. Even readers who sympathized with the work's intellectual direction found it difficult to follow the long book to the end. According to friends, it was like a visit with him. The great scholar would begin a wonderful, unforgettable monologue, but could hardly sit still and kept rushing over to a shelf or his desk to grab a book or object to clarify a point, switching apparently unconsciously between several languages, conducting disputes with colleagues, citing historical events, reading out new publications or reciting poems by his esteemed colleague Goethe.

The great man left his audience bewildered and a little intimidated, and only the bravest and most inquisitive continued their conversations – this too was an aspect of the natural selection described by Humboldt's friend and contemporary Darwin. When Humboldt managed to focus his ever restless attention, however, he could make his point with great eloquence:

Nature considered rationally [...] is a unity in diversity of phenomena; harmony, blending together all created things, however dissimilar in form and attributes; one great whole animated by the breath of life. The most important result of a rational inquiry into nature is, therefore, to establish the unity and harmony of this stupendous mass of force and matter [...]. Thus [...] is it permitted to man [...] to comprehend nature, to lift the veil that shrouds her phenomena, and, as it were, submit the results of observation to the test of reason and of intellect.[15]

Humboldt too was seeking to control empirical observations, but not nature itself. He had experienced nature as too huge, too changeable and, above all, too inseparably intertwined for anyone ever to dominate it.

Entangled Life

Alexander von Humboldt was building on a tradition that extended from the atomists of ancient Greece and the Roman poet Lucretius to Montaigne, Spinoza and Diderot, and had set itself the goal of describing the perceptible universe empirically on the one hand, and, on the other, using this perception to form theories that revealed the knowable world in its connectedness and strangeness.

From the mid nineteenth century, however, the development of science led to a point at which physics terminated its pact with perception. Even Humboldt's rational consideration had to capitulate in the face of relativity theory and quantum mechanics, since they contradicted every empirical reality, but simultaneously had an immense ability to make predictions and describe states. They worked as scientific theories, but they had left behind what humans could see and feel. No one could ever study an object travelling near the speed of light or locate a mass particle with their own eyes, but the possibility of thinking systematically about these phenomena and calculating them nonetheless changed the world and the way humans understood it.

Relativity theory and quantum mechanics contradict each other in certain aspects that have still not been theoretically resolved, but they have one decisive thing in common: what makes them stand out is their insistence on context, on the impossibility and senselessness of describing an isolated individual, object or event without surroundings or relations. In these mathematical models, physical phenomena are conceivable only in relation to others, only receiving their identity from their location. Space and time had to exist as a continuum, and any reflection on the one or the other was nonsensical and only permissible because human life took place at such low speeds, and in such a small space, that the resulting inaccuracies were of no consequence.

Quantum physics criticized not only its own discipline, but also the language itself. It made no sense to speak of objects with a particular location and a particular impulse existing as an independent reality. On top of that, the mere act of observation affected the events, and this too was not fully measurable but existed only as a statistical probability.

The Italian physicist Carlo Rovelli has called quantum physics a theory with extraordinary explanatory power and effectiveness (without it there could be no computers, internet, lasers ...), but with one blemish: it makes no sense. It forces us to think in paradoxes and accept contradictions in order to arrive at verifiable results. In other words, one of the most important and oft-proved scientific theories actually undermines the scientific method.

'I should only believe in a God who knew how to dance',[16] says Nietzsche's Zarathustra, and if this God is the God of Spinoza – God or nature – then he really does dance, for one always dances with someone and this dance is one that only a God could dance: a dance of all molecules with all elementary particles, a wild, ecstatic, anarchic undulation of material in forms and constellations that scientists in a variety of fields only began to discover a few years or decades ago.

On the one hand, this wave of new insights comes from the development of new technologies and measuring tools, as well as the use of Big Data and AI; on the other hand, it also comes from the simple fact that the convergence of different scientific developments has caused new questions to be asked.

No, not to worry: we are not moving towards esotericism, spirituality, homeopathy or subtle bodies. Our concern here is science, theory building through hypotheses and experiments, exploration and knowledge of the material world, even if that world dissolves on closer inspection into energy states – though those too can be scientifically examined – or is obscured by ever more complex mathematical models.

The scientific disciplines that discovered shared horizons here were biology and cybernetics, complexity research and climate physics, anthropology and game theory, behavioural biology, microbe research and botany. The results of these strange cognitive interferences across disciplines were astounding. A natural world that had been measured, evaluated and captured under the microscope suddenly began to speak in new languages, almost as if someone had finally asked it an intelligent question.

One example that became quite well known through the English biologist and mycologist (fungus expert) Merlin Sheldrake is the life of a forest. Sheldrake was not primarily interested in trees, plants or animals, but in the types of fungus that grow in forests – or, more precisely, their roots, known as mycelia. Fungi are fascinating organisms, and it had long been known that the mushrooms hunted by aficionados in late summer are only their fruits. The actual fungus – not an animal or plant, but a life form of its own – proliferates below ground in the form of a gigantic network of microscopically fine roots that grow as one organism without a centre, seemingly making intelligent decisions independently of one another as to whether they will continue growing in a particular direction or move their activities in a different direction.

Merlin Sheldrake is still a young man, and his field of research was only recently established. What he and his colleagues found out has revolutionized the scientific understanding of natural systems. Mycelia not only form the root fabric and the true bodies of the fungi – they are interconnected with the trees of the forest. With its enzymes, the mycelium extracts minerals from the soil and makes them available to the tree roots while the tree supplies the mycelium with sugar. This basic symbiosis is only the beginning of a far more complex relationship that has barely been understood. The mycelium also allows trees to communicate, for example by warning one another of infestations so that other individuals can develop antibodies; beyond this, the fungus roots even allow individual trees to supply nutrients specifically to other individuals, such as young or damaged saplings.

This fungal network has meanwhile been given the nickname 'Wood Wide Web' because it genuinely seems to function as a communications network within and even between forests, and the forest, which science had viewed until then merely as a collection of trees, gains a completely new character as a communicating organism that acts out of solidarity (and occasionally antagonism) and is gifted with its own form of strategic intelligence.

It is insufficient, then, for natural science to analyse a tree based on the most varied parameters, as one might analyse a cog or screw, then multiply these results with the number of trees in a forest to form an idea of how the forest lives and functions. On the one hand, each tree is already a forest: a space of communication between different organisms, a place of asylum and existential struggle for microbes and mites, viruses and bacteria. On the other hand, the forest is a cosmos of such microcosms, a symbiotic organism that gathers all manner of species in and around itself, a system of stupendous complexity in which trees rely on fungal roots to communicate among themselves, supply one another with information and even nutrition, react to pests together and act strategically: a constant interaction between animate and inanimate actors, from minerals and microbes via mycelium to all flora and fauna.

Knowledge that people who were closely connected to natural habitats have always passed on has now become a field of scientific research, even if theories and methods still need to be developed in order to understand the full scope of these discoveries – especially since the findings of the mycologists can also be extended to other areas of knowledge. Watercourses and mountain landscapes, oceans, forests and steppes, deserts and seabeds form systems that can best be read and understood – for natural scientists too – as actors in a cosmic activity of innumerable communication lines, dependencies, symbioses. Each of these systems seems to concentrate cascades of subordinate and ever smaller systems into a functional resonance field while itself becoming part of ever greater and more complex systems: a flowing and swirling on every level, from elementary particles to the water from the tap and immeasurable, distant galaxies – a cosmic dance of creation and destruction. This dance is danced neither for nor against humans; it simply takes place, like a storm.

A further new continent of natural science awaiting exploration, where communication and cooperation lead to new forms of existence, is the discovery of the (human) microbiome. It has been known since as early as the twentieth century that the digestive tract of all animals, and hence also humans, contains a particularly large number of microbes, which are important for the chemical breakdown and absorption of nutrition. The volume of the human microbiome is roughly that of a cup. Researchers had cultivated individual stems of these microbes in laboratories, observed and documented their metabolic processes. Soon it seemed that everything important had been said about this phenomenon.

Only in recent decades, thanks to better instruments, techniques and genetic analyses, has a new dynamism entered this field. The microbiome, which is the ensemble of all extraneous micro-organisms living in and on the body, is not only immensely larger and more diverse than formerly assumed; demographic-medical studies also show that it has a fundamental effect on every aspect of human existence. Meanwhile, large and well-funded initiatives such as the Human Microbiome Project (founded in 2008) are examining the genetic sequencing and character-istics of the human microbiome.

In keeping with good scientific practice, scientists who occupied themselves with this subject in the past isolated individual organisms and cultivated them in pure cultures. Their scientific socialization, their laboratory protocol and perhaps also the quantity of data they could assess prevented them from examining what happens when these species are no longer kept in pure cultures, but can communicate and react to one another – and when there are not dozens but tens of thousands of different species and organisms that split into countless billions of individuals, which all grow and digest, convert chemicals and produce others, which in turn create new biological surroundings and constitute a sort of biological documentation of collective experi-ences and environments. For this creates a completely new landscape of biological complexity, a landscape in which humans in particular take up an entirely new position.

According to the current state of research (which is a dynamic field), a human body harbours just under 40 trillion bacteria, fungi and viruses, substantially more than the body's own cells (which number roughly 30 trillion in an adult). These individuals belong to thousands of species and add a collective genome of 2 million genes to the human genome with its 20,000 genes; each human body thus contains 100 times more non-human than human genetic material.

In purely quantitative terms, the human body is a transport system and a source of nutrition for microbial life, but this is no passive coexistence. Recently published studies connect the health and diversity of microbiotic activity not only to the development of carcinomas, nutri-tional intolerances, allergies and diabetes, but also to the likelihood of having autism, senile dementia, Parkinson's disease or clinical depression. The microbiome does not help humans digest; it constitutes them all the way to their consciousness and perceptions.

After the decoding of the human genome by the Human Genome Project (1990–May 2021), increasingly precise analytical methods also revolutionized genetics and enabled a new understanding of how life is organized and how information is passed on. Along with the new continent of the genome, a revolution within genetics also changed the understanding of heredity in a way that completely overturned the image of the self-enclosed individual making free choices in the light of reason: what became known as epigenetics.

Epigenetics is a branch of genetics that formed around 1990 and studies the heredity of phenotypic traits without changes in the DNA sequence. Less academically put, the function of individual genes can be influenced, suppressed or reinforced through chemical markers that attach themselves to the genome. These chemical markers, which attach themselves to the genome, correspond to strong experiences such as famine, physical pain or historical trauma in individuals and can be passed on from one generation to the next. Even more simply put, an echo of historical experiences can be inherited, at least as a chemical marker that can still influence the functions of cells and messenger substances, diseases and even instinctive reactions in a subsequent generation. Epigenetic inheritance mechanisms for behavioural patterns have been found in experiments with animals, even when there was no contact between the parent animals and their young. In humans too, studies have shown that traumas in one generation can affect the expression of genetic characteristics in subsequent generations.

The discovery of epigenetics means that the human genome is precisely not a rigid, mechanistic blueprint used to construct an individual exactly point for point, but that these instructions are more dynamic and alterable, that they react to influences from their environment and are shaped by the experiences of earlier generations, even down to the metabolic processes of individual cells.

The image of the genome as an exact set of building instructions, which went well with the metaphor of humans as machines, proved to be misleading. Perhaps this is a case of life imitating art, not vice versa, for the way the genome works resembles what is also demonstrated by the 'cultural DNA' of a society: it consists of information – narratives and memories, traumas, attitudes and rituals – that is passed on from one generation to the next. In this transmission, however, mutations occur not only as chance errors, but also through the integration of experiences, migration, new stories that grow on top of the old structures and can offer entirely new and different functions. This is a cultural image of how the traumas of long-gone generations literally live on in the cells of their descendants, where they continue to be present at the level of bodily functions and hormonal disturbances.

Both the examination of the microbiome and that of epigenetics are still in the early stages, and the findings are constantly being revised, but it is already clear that the scientific exploration of complex microbial and chemical communication processes will lead not only to new possibilities of medical diagnosis and methods of treatment, but also to a new conception of humans.

All of this does what good science is supposed to do, for it transcends its own frame of reference not only in methodological terms, but also metaphysically and in all its political implications, which are hardly insignificant. How do these factors influence accountability in criminal trials or equity of distribution in a society, equal opportunities, the very possibility of justice? Is this a variation on the traditional tale of a rough childhood, or does it point to the fact that there is a biological basis for different experiential horizons and a different scope of action?

Arguments of this kind risk drifting into biologism, but they touch on a deep aporia in democratic societies, on the unanswerable questions that defy all models and explanations. In the tradition of the Enlightenment, which certainly has a theological character, they proceed from the assumption that citizens are free, rational individuals who are willing and able to inform and involve themselves responsibly, at least to the extent of voting.

Scientific insights outline an entirely different being: extremely susceptible to manipulation and full of cognitive prejudices, never truly in control of itself, inhabited by a consciousness whose disposition is a result of microbial activities and whose bodily functionality rests on a series of unknown factors – a being that, according to cognitive analyses, displays severe intellectual deficiencies and problematic tendencies.

At this point, philosophical questions prove deeply political and potentially revolutionary. Will the idealistic vision of the democratic order be marginalized in the same way as the theological explanation of the world that preceded it? How can liberal societies deal with a changing scientific data set that reveals humans as more irrational,

vulnerable and manipulable than democracy would – and must
– suppose?

This is not only about questions of political philosophy, however,
which can be countered if needed by pointing out that, even as necessary
fictions, the self-responsibility and free will of the bourgeois universe
still serve the purpose of maintaining the public order, even if not every
single judgement reached in this context is just.

The scientific reinterpretation of humans unsettled an entire tradition
of thought. The old-style lord of the creation evoked by the Neoplatonists,
the Bible and the Abrahamic traditions stands above nature and subju-
gates it to his free will. He is physically isolated from the outside world
and intellectually sovereign, master of his own actions (the consistently
masculine characterization is no coincidence), and his body is created in
the image of God.

In contrast with this familiar image, the human being is a profoundly
symbiotic organism, a product of unknown communication processes
between countless life forms whose activity influences or even dominates
its entire life and experience, down to its very emotional disposition,
intelligence and perception – an organism whose experiences are passed
on across generations, and whose individuality must accordingly be
conceived in a different way. This paints a shocking and unexpected, yet
also deeply liberating, picture of the primate *Homo sapiens*, which is just
in the process of taking a huge epistemic step. The sovereign human being
as a carrier of the soul fades away, and a wilder, more entangled and largely
unknown creature shows itself beneath the mask of theological tradition.

The long-familiar type of human being, first described theologically
and then, at the turn of the millennium, economically, is replaced by a
phenomenon that we may still lack suitable words to describe after 6,000
years of linguistic development: a form of experience, a consciousness,
a horizon of desire and pleasure and pain, an aspect of an immense
symbiosis between many thousands of species, an unprecedented chatter
and pulsation of messenger substances and electric impulses through
the pathways of a body that constantly exchanges and renews itself,
producing 100 million red blood cells every minute. Molecules from the
world around it are incessantly incorporated, transformed and excreted;
only the experiential self that rules over this fantastic activity remains
stable.

The human being as a sovereign rational actor is exposed as last season's model. The old Adam is replaced by a symbiotic biological event horizon with its own internal puppet theatre. Like a single tree in a forest, this organism cannot be fully described or understood as an individual, but only as a part, a nodal point in a network of events and motivations, a momentary state of the life that vibrates through all these states.

None of this makes the puppet theatre, our individual experience – which is all we have as humans – less important, less fascinating or less decisive as criteria for what a good life might be, but the reversal of perspective from the perspective of natural science calls to mind Montaigne, who once asked in bewilderment how such a creature 'who is not even master of himself [...] should call himself master and emperor of the universe'.

The scientific-cognitive revolution currently taking place proceeds from the repeatedly empirically confirmed assumption that natural phenomena cannot be adequately described through an analysis of their parts, or of separate individuals. Instead of seeing fixed, stable objects, it thinks in terms of communicative acts, events and connections that draw their identity and potential for action from their context, temporarily assuming certain states and functions where any change is reflected in the transformation of the whole.

As an analytical approach, however, this perspective on so-called nature developed from scientific models seems much stronger as a way of describing phenomena and interactions than the preceding mechanical image. Only in this way is it possible to move closer to the infinite complexity of natural systems and better understanding of how they function. The condition for this improved understanding, however, is to relinquish the special status of humans and, with a certain epistemic modesty, to define the possible place of *Homo sapiens* within these contexts. The price of this revolution is radically thinking the human being as inescapably entangled with the existence of other living and non-living actors on this planet.

A Handful of Earth

At least since the early twenty-first century, it has become clear that the human project of subjugating nature was a disastrous mistake. It is thwarted by ecological reality and chokes on its unwanted side effects. Paradoxically, the project of subjugation was doomed to failure because the technological scope of human insatiability was suddenly expanded immensely by fossil fuels, and the internal dynamics of growth processes developed so rapidly that there was no time left to understand this transformation and reflect on its dizzying development, and, with the energy of the ascent, people probably felt no need for criticism.

During these immense transformations, which affected almost everyone on the planet in some way and caused the total population to grow to more than double within six decades, more brutal subjugation, more efficient exploitation and increasing growth were successful strategies on a global market. With the first effects of the climate catastrophe, however, this logic collapsed.

In the post-war years (and really since the sixteenth century), Western governments could only come up with one answer to very different structural problems: economic growth. But this answer was made obsolete by the climate catastrophe. More growth, more exploitation and more dominance of human interests no longer lead to prosperity, freedom, security or control, but increasingly to uncontrollable developments, the rise of sea levels and changes in entire weather systems, natural disasters such as hurricanes, heatwaves, floods, desertification and drought, and ultimately the migration of millions of humans and other life forms in search of survival.

The increasingly evident consequences of the climate catastrophe prove that a subjugation of nature is unfeasible, partly because it rests on the Bronze Age idea that humans stand outside and above 'nature', and can safely exploit and change it without being affected themselves. With the increasing economic, human and ecological costs of the system changes set in motion by human interventions, the certainty also increases that the logic of subjugation no longer has any constructive strategies to offer that might make a difference in this situation. Every effect has an unwanted, unforeseeable side effect.

In more technologically primitive societies, the systemic narcissism of the lords of creation was one harmless illusion among many, but with the technological development of the twentieth century it developed into a suicidal delusion – or perhaps not even a concrete delusion, just a deep-seated sense of one's own inconsequential, context-free actions: a tacit social agreement that it is better not to talk about certain things or think them through too much.

From a purely natural-science perspective, the sudden leap in humanity's productivity and activity triggered a reaction that causes all human progress to falter, and even questions its existence, because old recipes suddenly no longer work – or have unacceptably strong secondary effects. A worldview that had for millennia provided the key to dominance and meeting challenges had no useful answer to this systemic dynamic.

The French philosopher Bruno Latour describes the reaction of an enlightened humanity whose history can no longer describe reality and whose model proves unsuitable:

> And yet – should we really be surprised? – the intrusion of the Earth strikes them as a shock. It appears that the Globe they expected to list, register, locate, enclose, and gobble up was no more than a very provisional rendering of what there remains to discover [...]. In the end, at the beginning of the twenty-first century, the Earth again appears – to the stupefaction of the rich enlightened portion of the human race – as *terra incognita*.[17]

But how do we navigate this unknown Earth? Latour, who came from a Burgundian family of vintners, sees the answer to the alienation of the Enlightenment in a return to Earth – not in a nostalgic or nationalist sense, but as a relocation of life to a concrete place, a physical experience of our own connectedness. There may, he reminds the reader, be more microbes living in a handful of earth than there are humans on this planet: a microbiome that runs through parts of the Earth's crust, a universe on which others build and which is linked to other systems.

As a good Burgundian, Latour grasps this not as the steaming clod of nationalism but as a *terroir*, the expression of the uniqueness of each place, every handful of earth that lends a wine its individual character, every entangled life.

Latour engages in depth with the Gaia hypothesis developed by James Lovelock and Lynn Margulis. Influential scientists long dismissed this theory as esoteric, and it was simultaneously appropriated by esotericists based on an equally significant misunderstanding, but it is very much grounded in natural science. It is a matter of understanding the Earth or its biosphere as an organism that attempts to create the ideal conditions for its survival.

Lovelock formulates his central idea as follows:

> Like coevolution, Gaia reflects the apartheid of Victorian biology and geology, but it goes much further. Gaia theory is about the evolution of a tightly coupled system whose constituents are the biota and their material environment, which comprises the atmosphere, the oceans, and the surface rocks. Self-regulation of important properties, such as climate and chemical composition, is seen as a consequence of this evolutionary process. Like living organisms and many closed loop self-regulating systems, it would be expected to show emergent properties; that is, the whole will be more than the sum of its parts.[18]

For Bruno Latour, the Gaia hypothesis offers an attractive possibility to rethink our existence on an interconnected Earth. In his conception, Gaia is simultaneously a metaphor with ambivalent, quasi-religious resonances and a cognitive status description of the unknown continent that arises from the shift of perspective in contemporary science and its models.

The idea that Latour develops in his writings, but also in public exhibitions and in multimedia performances, adds an icon to both the scientific and the political imagination. Instead of talking about the fact that people live 'on the Earth' – that is, metaphorically with their feet in the dirt but their heads held high and their eyes directed at distant horizons, as if to underline their elevation above nature – Latour suggests thinking about humans as inhabitants of the 'critical zone': the thin membrane, visualized by Lovelock, between the dead rocks under our feet and the eternal emptiness above our heads, the only environment in which life is actually possible.

> At the scale of the usual planetary view, the thin surface of the critical zone is barely visible, it being only a few kilometres up and a few kilometres down at most. It is no more than a varnish, a thin mat, a film, a biofilm. And yet, pending the discovery and contact with other worlds, it is the only site that living beings have ever experienced. It is the totality of our limited world. We have to imagine it as a skin, the skin of the Earth, sensitive, complex, ticklish, reactive. That's where we all live – cells, plants, bugs, beasts and people.[19]

For Latour, this rethinking comes at a high price. The people of the modern world begin to realize: 'The present crisis is that progressively the Moderns realize they can no longer count on this vast and unlimited reserve for their future prosperity. Modernization, development, globalization appear to have been a fortunate and undeserved parenthesis that is now coming to a close.'[20]

The challenge is to radically expand the concept of agency, the potential for action, for the climate catastrophe also makes non-human actors part of a process that, from a human perspective, is essentially political and ecological, revolutionizes the philosophical conception of humans, and calls for new practices and a new form of attention:

> It might no longer be defined as a collection of humans exploiting the resources of a soil through a system of production, but rather the many contradictory and disputed assemblages of all sorts of entangled life forms striving to persist in time and expand in space a bit more. The old distinction between society and nature is being replaced by an agonizing process of composition among agents – humans and non-humans – all clamouring for recognition.[21]

Here, in wonderfully ironic fashion, a historical circle closes. One element of humanity's experience in the climate catastrophe is the fact that it cannot possibly continue to act as if human history, the human consciousness and the role of humans were separate from nature, as if they were not part of natural processes and cycles, constantly involved in complex, uncontrollable and opaque interdependencies that can be glossed over but stubbornly continue to exist nonetheless, and whose effects cause massive destruction.

This realization that the life and survival of humans and entire societies are radically dependent on natural processes, and that a successful existence must be based on a carefully calibrated give and take, brings us back to the age of Gilgamesh, for the science-based worldview has a surprising amount in common with ancient, polytheistic conceptions of nature.

The average Sumerian or Akkadian of the second millennium BCE had a relationship with their gods that only differs in details from those of a Māori, an Aztec, a Tibetan or a woman in classical Athens. They all knew that their environment was inhabited by gods and spirits, ancestors and demons, nymphs or furies, and that the world around them was pervaded by these interests and their actions. They also knew that they themselves were inhabited and driven by forces they barely understood, and that it was important to employ the best possible techniques and means of understanding to enter into mutually beneficial relationships with these forces.

Every action made it clear that they were embedded in a web of powers and interests that had to be acknowledged. Every action was an intervention in the existing fragile balance, and interfered with the orbit of another form of existence. Someone who wanted to set sail had to sacrifice to the god of the sea; someone planning a journey asked the oracle about the will of the gods; someone planting a field built an altar to the spirits living there, which were as present as the ancestors and the gods themselves. Every human act was embedded in a depth that extended to myth, an infinitely complex web of magical forces and heroic tales, divine intervention, gifts and counter-gifts – a constant give and take between humans and the world they inhabited. Whoever failed to do this, like Gilgamesh, would suffer the consequences of their hubris.

The metaphorical language of animism poetically expressed the mutual entanglement of human and non-human interests and actions as a connection to the ancestors, spirits and gods, and this relationship was formalized through rituals and sacrifices. This was more successful at the metaphorical level than the scientific, for even when Poseidon had graciously accepted an expensive burnt offering, a storm might still smash a ship and crew against a rock along with their precious cargo. However, there was a cultural framework in which giving and taking, in connection with natural events such as birth and death or sowing and harvest, were symbolically and ritually acknowledged and fulfilled.

Here, once again, we see how metaphysically radical the idea of subjugation actually is, since it constitutes a break with a complex understanding of the world based on mutual dependencies. This worldview was shattered with the command to Adam and Eve to subdue the Earth. From now on, the relationship of humans was no longer based on a give and take; it was not reciprocal at all, but rather consisted in humans 'conquering' and 'taming' nature, 'overcoming' mountains and 'discovering' continents. Such terms have clear sexual connotations: man pulls back the cover from the intimacy of the land in order to possess it, which may include penetrating it with a flagpole. The Earth itself, 'nature', was passive and no longer had a voice; it could be possessed and sold, penetrated and cultivated, exploited and dominated by God's beloved but fallen creature, the human being.

Now the climate catastrophe is posing the compelling question of how far the insight of animist cultures into the interdependence of human and non-human actions and desires was actually appropriate for the real relationship between *Homo sapiens* and the vast majority of an infinitely complex nature, even if the idiom in which this knowledge was expressed is not quite suited to the present. Where myth speaks of gods and goddesses, natural science knows its own systemic thinking with its own actors, perhaps even its own scientific rituals.

Metaphorical languages change, but the fundamental assessment of the relationship between humans and the world that surrounds, pervades and constitutes them was perhaps more realistic than the millennia-long foray into narcissism, which revealed itself in the early twenty-first century as an existential error.

Once the grand scenery of its subjugation history has been taken down, however, a naked ape is left on the stage. The last play that was performed here, *The Free Market*, forced the ape to contort itself immeasurably as a successful, happily consuming individual and work itself into a state of permanent nervous distraction. But this play too has come to an end. Perhaps it is time to think about a species-appropriate way of keeping *Homo sapiens* that is devoid of cruelty. Even Hagenbeck's ethnographic exhibitions made more of an effort in this respect than the sinister prophets of the digital future.

Risky Thinking

Let us take a risk. If we take the Enlightenment idea further, based on Kant's principle of humans emerging from their immaturity, what would that look like? And do we even want it?

The Enlightenment's first aim was to place knowledge of the world on a solid foundation, empirically and scientifically, using only reason and verifiable perception; its second was to argue for a society in which power rests not on tyranny or tradition, but on the free and equal decisions of reasonable individuals and an idea of common knowledge, of progress.

In truth, this basic description raises more questions than it answers, and juggles with concepts that can be defined in completely different ways – especially that of reason. That is what makes Kant's essentially poetic, non-specific suggestion of emerging from immaturity so strong: he does not involve himself in the game of definitions.

However, it is actually important for concepts to be sounded out and debated over and over, for there have been many political or ideological declarations in the name of the Enlightenment, but little intellectual seriousness. *Liberté–égalité–fraternité* makes a wonderful battle cry for the idealistically inclined, but how does one flesh out this freedom when one person's freedom always means another's unfreedom? What can equality mean when it comes into conflict with freedom? Who is included in inequality, who is not and why? Does the idea of fraternity not reveal the patriarchal structure of the whole edifice, which has always oppressed women and in which oppression is the aim? As soon as one starts to unpack and question these words even a little, they lose their clarity. Perhaps it would be wise to find an approach to these definitions inspired by Heisenberg: it is possible to make statements about fields of meanings, but not to pin down every individual particle in each field.

These debates about definitions have broken out again in recent decades in the shadow of feminism, postcolonialism and gender studies, and have contributed enormously to deepening these questions and expanding reference points. So new life has been breathed into the old Enlightenment debate, and in this context it seems useful to reassess the sustainability of the Enlightenment's original ambition.

The Enlightenment was the attempt to think humans within nature. Early Enlightenment thinkers such as Spinoza, Bacon and Descartes (and before them, as a solitary voice, Montaigne) set themselves the goal of basing their knowledge only on what they could observe and logically conclude from their observations of nature. Descartes failed because he simultaneously wanted to include an almighty and benevolent God in his system, which forced him to make various hair-raising compromises, such as the claim that only humans had a soul and accordingly a personality, feelings and memories, while animals were no more than biological automata.

Such compromises were made by many thinkers over the following two centuries. They tried in various ways to use the impetus of the Enlightenment movement to pursue their own social and economic interests, but also to make peace with theology and not go out on too much of a limb.

What is generally referred to today as 'the Enlightenment' corresponds, as already mentioned, to Jonathan Israel's 'moderate mainstream': the authors whose thinking consciously or unconsciously continued a Christian theological tradition, albeit with a different vocabulary: salvation history became progress, the soul became reason, humans were outside and above nature and they had the special task of subduing the Earth. At the same time, a professional justification industry ensured that enlightened minds could construct their imperialist campaigns as moral necessity and their wealth as personal sacrifice.

In the logic of the moderate Enlightenment, this tendency to project power outwards corresponds to an economy of emotion infused with a profound Christian hatred of the body, as well as an inward projection of power. Enlightened morality often viewed naked lust with the same prudish eyes as the church fathers. Desire did not fit into the conception of a being driven by reason which controls itself so perfectly that it can make free, rational decisions in discourse with equals.

Without the theological ideas transported and, in a sense, rebranded by the moderate Enlightenment, *Homo sapiens* seems strangely and very unusually naked. He or she stands in the midst of nature, a primate species whose astonishing successes now threaten its existence, which is actually an old evolutionary story. This primate species does not behave any more rationally than other animals – its life goals and social strategies, its fears and longings, are similar, but this species has a more complex language, which enables it to channel its destructive emotions into rational-sounding statements and conceal its true motives with more elaborate lies than other animals can.

This *Homo sapiens* cannot subjugate nature; it cannot, as Montaigne pointed out, even understand its own nature, let alone control it, and therefore becomes entangled in endless emotional proxy battles. In truth, its experience is a temporary event that desperately and obstinately postulates its own eternity: a primate that takes offence at itself because it has an idea of itself that it cannot fulfil.

None of this reduces the existential dramas or great dreams of individuals and entire societies; it simply disabuses them of the absurd assumption that those have any significance *sub specie aeternitatis*. The history of humanity will end not with a solution to all problems and the arrival of perpetual peace, a utopia of definitive justice or even a dystopian rule of evil; it will end more chaotically and, for very prosaic reasons, without a spectacular finale.

The finitude of *Homo sapiens* is only a tragedy if one is still under the spell of the progress narrative, if the idea of a divine plan is still lurking somewhere in the woodwork of one's conception of history, occasionally making its presence known through loud creaking sounds or the dust of centuries trickling down. Animals are mortal; so are civilizations and species, and *Homo sapiens* is no exception. Nonetheless, there is nothing more important than shaping the future possibilities of the life we are currently living in the best possible way; humans have an ethical relationship with their descendants, just as they do with their ancestors. Perhaps that is not very enlightened, but it is an anthropological constant.

Besides giving an overview of the dominant Western tradition of the last three centuries, the modern Enlightenment and its highly efficient recycling of theological ideas, I have repeatedly tried to make room for other perspectives within that same Western history of subjugation, for there have always been people who had the desire and the courage to follow the implications of enlightening energy with more will to radicalism. In these works and debates, the world often appears foreign and cold, a human merely a collection of atoms living in a continuum of atoms, nodal points in an eternal process of transformation, a mysterious life and experience of life that becomes central to all human endeavours. Time and again, these thinkers remarked – with a profound, almost mystical respect – that with such ideas, they were ultimately describing a world whose true workings were still entirely unimaginable to them, a physical reality for which they still lacked suitable models and means of understanding.

Since then, the possibilities for understanding science have been revolutionized, but this has only deepened the human dilemma. Human imagination and perception are already helpless in the face of modern cosmology or quantum mechanics. The human drama within this increasingly foreign universe, however, along with the horizons of our emotional understanding, has remained the same.

With its ever improving, albeit inevitably schematic, understanding of the world, *Homo sapiens* is cast into a foreign reality. The Enlightenment was the beginning of this path, but was also strongly influenced by a fear of facing the full implications. The time has come to think these implications too, since it has become an existential threat to continue ignoring them.

The imaginaries and images of a culture grow with and through one another. The Enlightenment detached the images of a creation, replacing them with a mechanical conception of the world based on reason, an idea that went perfectly with the logic of manufacturing, capitalism, industrialization, empirical science, colonialism and the perfecting of society as a panopticon, a eugenic experiment or a transhumanist future.

With the interconnection revolution, a new image took centre stage. This image too is only a crutch for our understanding of a fundamentally alien reality, not the ultimate truth about it – and this image too has come about, and is still coming about, in a disordered dialogue with the society from which it emerges, and which it in turn shapes. In a world of global interconnection and dependency in which social, sexual and cultural identities seem increasingly insecure, contingent and artificial, this image of natural connections corresponds to the emotional attunement of society.

It is a thankless task to dismantle the theological scenery of the Enlightenment and leave the protagonists standing on stage in surroundings in which they have no orientation. But when the assumptions about the position of humans in the world begin to unravel, so does the enlightened vision of the future. A fundamental failure of the Enlightenment that was pointed out by Karl Marx was its linear conception of the future.

An affectionately illustrated article in the *Encyclopédie of Diderot and d'Alembert*, 'pompe à feu' (fire pump), describes a peculiar machine that pumps water and is driven by fire. Writing in the second half of the eighteenth century, the authors describe an early steam engine without realizing that they are looking at a technology that will radically transform not only industry, but also economies and societies. They are developing their utopias linearly from the eighteenth century. Their utopian societies are still based on agriculture and only know the community of equals (although certain people are occasionally more equal than others), in which, once the perversions of the church and the monarchy have been overcome, humans will decide and live reasonably and virtuously.

What makes this enlightened utopia so bloodless is that the people who inhabit it are theological constructs in the same way as the people in neoclassical economics. Thus, a question that was asked too rarely in the course of the Enlightenment still remains: what if not all of a country's citizens want to use the privileges of their freedom and equality? What if it doesn't matter to them who holds power and what they are doing to other people, as long as they themselves have enough food and entertainment? What happens to the Enlightenment if the Romans were right about *panem et circenses*?

This last question is often dismissed with moral emphasis, even outrage, as if it were unseemly or cynical to pose it. But after three centuries of Enlightenment and several generations of democracy, with a degree of freedom that is historically unparalleled, the ethical and political ideas of the Enlightenment thinkers have not been realized in any way, even in rich societies. Yet many of them were quite convinced that all people would educate and inform themselves, participate in political decision-making and make an effort to improve their society if they were only given the chance. That this is not the case, even in largely free and democratic societies, indicates that the conception of humans among those thinkers was based more on their own ideals than the empiricism they so often invoked.

This is one of the central battlegrounds for the scientific exploration of human behaviour and its potentials. Are humans – beyond small groups, families or clans – capable of living and working together with strangers, with people they find hard to read, even trusting them and investing their trust and security in institutions designed to protect them? Or were the democratic experiments of the twentieth century no more than a side effect of the oil boom, impossible to repeat without immense economic growth? Are people in the plural not only able but also willing to become the individuals the Enlightenment believed them to be, or was that old cynic Voltaire right when he suggested having one truth for the enlightened and another to keep the rabble under control? Is it enough to keep society fed, entertained and distracted in order to exercise power? What does the new moral, cognitive and ethical discovery of the unknown network event *Homo sapiens* mean for the primate and the citizen?

Are humans in societies whose size and complexity go beyond individual social circles, a particular class or a particular type of person capable of having sufficient trust, showing real solidarity and cooperating with the institutions of the collective, instead of seeking dominance or regressing to tribal reflexes? Are they capable of envisaging themselves without subjugation?

The range of human social behaviour oscillates roughly between the relaxed free love communes of the bonobos and the martial hierarchies of the chimpanzees, with whom we share around 98.5 per cent of our genetic material – that is, the genetic difference between us, chimpanzees and bonobos is the same as between Indian and African elephants. What is the potential of *Homo sapiens* as a social being? Only history can reveal that.

Let us work with the hypothesis, then, that the emergence from immaturity is more psychologically complicated than Kant and his colleagues believed, but that it is perhaps not impossible – not entirely, and certainly not at all times and in all places; but there must be a perspective beyond this emergence leading to a landscape that can become navigable.

We think about nature in mental images: as creation, as a machine, as a critical zone, as a self-regulating organism. None of these images is true, correct or complete, since each of them uses its own metaphorical language and can describe certain experiential aspects in a profound and detailed fashion, while others remain shallow and ill defined. This is in the nature of such an image, and Francis Bacon and Claude Lévi-Strauss knew that humans need totems in order to reflect on the world around them.

The decisive question to pose with an image of the world is not whether it is true, but whether it can lead to constructive action in this world. Will someone reach their goal with a particular map in their head? And is it a goal worth seeking? It only makes sense to ask whether the goal on the map is worth the effort of reaching if the map describes the landscape sufficiently well. If the map describes the world as an asymmetry between rulers and ruled, then all paths lead through the terrain of oppression. If a map uses borders to depict the territory, a viewer thinks about it differently. If the map shows connections, dependencies and communication processes, it inspires other ways of moving through the world.

In a sense, this history of subjugating nature has also sought to contribute to the drawing of such a map, one that thinks in connections rather than boundaries, in cumulative facets of meaning rather than definitions. This way of thinking is slightly reminiscent of the new view of trees as part of a communicating biological organism. The clear definition of a tree as an object surrounded by an imaginary black line makes no sense in this context. The tree can be better understood as a communicative event. Its trunk is solid and outwardly clearly defined, but the branches, the twigs and especially the roots spread out, merging functionally and communicatively with the root suckers of other trees and the mycelium of fungi and other organisms, such that the tree extends even beyond the end of its roots. It is only from the perspective of this interaction that a meaningful image of the tree comes about.

From this perspective, a river can also be understood in a different way: not as a demarcation between the countries at each shore, but as a transport route, a space for life cycles that often take place over hundreds of kilometres along the river's course. The river connects and transports an almost incomprehensible and still largely unexplored multitude of substances, information, animal and plant species, microbes and stories, enabling them to create new life contexts and new complexity, new forms of unstable equilibrium.

Thinking the Enlightenment further would lead to a multidimensional, ever changing map of connections and entanglements that recognizes borders, but treats them as one of the less interesting aspects of a landscape of dependencies and conversations. Such an Enlightenment would search for complexity in the simple, connections in the isolated, swarming life where it is not obvious, the uniqueness of experience: the *terroir* of existence. Its symbol would be a handful of earth.

Thinking the Enlightenment further would mean radically rethinking the place of humans as one element in a nature that is no longer a subjugated Earth, but rather an infinitely interconnected, interdependent system of systems that blur boundaries and other scientific categories, narratives and images, and require other artistic interventions and personal experiences in order to become tangible. This conception of nature involves an interpenetration of the individual and the collective, the living and the non-living and cause and consequence in a way that can be mathematically expressed, but not imagined. This is truly a *terra incognita*.

Thinking the Enlightenment further would begin with examining one's own ethos in the Foucauldian sense, examining the attitude and assumptions of one's culture to find structures that make rational, clear thinking more difficult by guiding arguments in a certain direction and adding a certain valuation, a peculiar colour, a smell or a baggage to their content. Every word bears the imprint of its history, but they are not all as tendentious and intangible as that endlessly charged word 'nature'.

It is a constant challenge to identify in the structures of supposedly secular historical, philosophical, scientific and political thought the core theological ideas that directed the gaze of Western tradition for centuries, and still exert a surprisingly strong influence – from the collapsing idea of subjugating nature to the corresponding conception of humans.

Ecology, Bruno Latour writes,

clearly is not the irruption of nature into the public space but rather the end of 'nature' as a concept that would allow us to sum up our relations to the world and pacify them.

What makes us ill – justifiably – is the sense that the Old Regime is coming to an end. The concept of 'nature' now appears as a truncated, simplified, exaggeratedly moralistic, excessively polemical, and prematurely political version of the otherness of the world from which we must protect ourselves if we are not to become collectively mad – alienated, let us say. To sum it up rather too quickly: for Westerners and those who have imitated them, 'nature' has made the world uninhabitable.[22]

How can the world be made philosophically inhabitable too? Is it possible to break out of the intellectual and linguistic division of reality and its logic of subjugation? This would be precisely in keeping with an emergence from our immaturity.

It is not a matter of language regulations, but of enabling resonance in Hartmut Rosa's sense.[23] In the course of history, nature has been described in various metaphorical images. The present book describes the mental images of a tradition in its approach to nature. These images or intellectual forms trigger resonances among one another, creating a field in which certain images and concepts can form the space of culture together. Certain images and words will develop a strong resonance, while others must first be introduced, or will inevitably stand out dissonantly from a constellation, because they are part of a different resonance field.

Could human societies and individuals come to terms with the world better, and draw a better map of it, if they could leave behind 'nature' and 'culture' as categories of difference to live in a world where such distinctions merely conceal the far more important fact of mutual entanglement and connection?

If it is possible to locate humanity in the midst of nature – not only scientifically, but also epistemically and existentially – and construct it anew from there, this would set off an immense philosophical revolution: the possibility of fulfilling the original promise of the Enlightenment after all.

D'Holbach's insistence that 'man is the work of nature' is also mirrored in the history of human thought. Perhaps the idea that humans can subjugate the Earth could only develop in a landscape like the one around Uruk; without the climatic shock of the Little Ice Age, the transformation of the seventeenth century and the debates of the Enlightenment, the *Lumières* and other national versions would surely not have come about. Today's climate catastrophe likewise creates the empirical necessity of rethinking the relationship between 'humans' and 'nature', though these concepts have grown increasingly fragile and full of holes in the course of this intellectual journey and can no longer hold any clear meaning.

The climate catastrophe changes our perception of the natural world and of our relationship with it. Like the pandemic, it is an unwelcome but also unmistakable reminder that the language of the occident and its heirs and imitators is no longer suitable to describe reality, and that this problem affects the root of language because it negates the worldview on which this language is based. In concrete terms, this means that the words that shape our thinking, the concepts that are ready to be filled with experiences, are still theologically charged, with all the cultural baggage that entails.

Wherever culture and nature, but also politics and global warming, or economy and ecology, are conceived of separately, wherever 'man' is elevated above 'the Earth', wherever history strives towards a paradisiacal goal or an apocalypse, wherever there is some great meaning that is supposed to underpin and guarantee everything, wherever 'humans' are simply different from 'animals', wherever one's own virtue always coincides with one's own benefits, and privileges are morally justified – wherever people argue and think in such ways, theological thinking is at work. Not in the sense that the speaker holds religious beliefs, but because these concepts immediately trigger a resonance in the other, a shared space, that goes back to the conception of the world in past societies.

However, this collective resonant space has long since become the cave in Plato's parable, the projection space of a shadow world that has established itself more firmly in the age of the smartphone than Plato would have imagined in his wildest dreams. In this sense, we are all Platonic idealists. The theological tradition no longer triggers any resonance in the natural world; it only obstructs our view of the endlessly fascinating connections between matter that only recently became measurable, describable, modellable and thinkable for the natural scientists – even though this intellectual process most probably leads towards a conception of the experiential world and its physical foundations of which we do not have the slightest inkling from the perspective of the present day.

It is entirely possible that systems of this complexity can only be theoretically conceptualized and explored via Artificial Intelligence and neuronal networks, and that the human brain is not actually capable of forming an intuitive, sensuous image of such a reality. The sensory organs and intelligence of *Homo sapiens* are adapted to perceiving a particular, tiny section of the physical reality that can be experienced by organic creatures, in which they can meaningfully act. Their imagination already fails when faced with large numbers and ideas that do not correspond to their experience.

At the same time, there is nothing more important than being able to capture reality figuratively and dramatically and translate it into appropriate acts. The psychological and cognitive crisis of the climate catastrophe is also expressed in the fact that a particular stance – that of the subjugator, which has prevailed for 3,000 years – has now turned against its owners, leaving an entire archipelago of societies without the established tools of comprehension.

What interpersonal resonant space can represent this new Earth? How can language, body, emotions and consciousness arrive on an Earth that must first be accessed linguistically because the old language has become worthless? And how can we get there from this old language? In the same way people always have: based on experience, as long it is not immediately forced into an enclosure of words and channels. 'Today we have to get started, we must decipher the narrative of the true conditions, just as people suddenly found the Rosetta Stone and were able to decipher the hieroglyphs. It is an art of translation. Collection. And keeping a grip on reality', as Alexander Kluge once said in an interview.[24]

This Rosetta Stone is still far from being deciphered, but the first letters and words are becoming clear – clear enough for us to say that nature speaks a language whose grammar has little to do with the sensory experience of *Homo sapiens* and the latter's mental constructions of its surroundings. The survival of humanity as civilization and as a species will depend on how well and how quickly humans learn to decode and speak this language.

Humans experience themselves (perhaps even more in the alienated environments of major cities than in traditional life contexts) as self-enclosed individuals, even if this is at best a traditional fiction from the perspective of natural science. Nature speaks a foreign language; perhaps that is precisely why humans developed their metaphorical universes in which they could tell other stories, stories in which they felt at home.

But can humans learn to feel at home in other stories? Can subjugation (including that of one's own self) be replaced by a different idea of the good life?

In the twenty-first century, humanity is being driven out of its conceptual homeland, the history of the subjugation of nature, as Adam and Eve were driven from paradise – except that this time it is not by any God, nor does an angel appear with a flaming sword, not even a wrathful Gaia. Rather, a predictable and long foreseen cascade of deportations within the critical zone, a change with potentially devastating feedback effects, descends upon them.

Let us not be misled by the parallel with Adam and Eve. This is not a moral story; there is no God behind it, no meaning or covenant or historical mission, not even the banal idea of progress. The expulsion from paradise without angels or salvation history is a painful experience, for it turns out that humans have an overwhelming desire for angels and salvation. They shiver in the new, strange nature, for they still lack the terms to describe it and have not yet understood that their expulsion is also their liberation.

Is this liberation really what they want?

Notes

Prologue: Buy Me a Cloud

1 www.theguardian.com/world/2021/dec/06/china-modified-the-weather-to
-create-clear-skies-for-political-celebration-study.

2 www.noajansma.com/buycloud.

3 Ibid.

4 Dipesh Chakrabarty, *The Climate of History in a Planetary Age* (Chicago and London: University of Chicago Press, 2021), p. 68 and *passim*.

5 Bruno Latour, *Facing Gaia: Eight Lectures on the New Climatic Regime*, trans. Catherine Porter (Cambridge: Polity, 2017), p. 15.

6 Ibid., p. 85.

7 Philippe Descola, *Beyond Nature and Culture*, trans. Janet Lloyd (Chicago and London: Chicago University Press, 2013), p. 30.

I Myth

1 *The Epic of Gilgamesh*, trans. Andrew George (London: Penguin Classics, 2000), p. 1.

2 *The Epic of Gilgamesh*, trans. N. K. Sandars (London: Penguin Classics, 1977), p. 61.

3 *Gilgamesh*, trans. George, p. 7.

4 Ibid., p. 70.

5 *Gilgamesh*, trans. Sandars, p. 102.

6 *Gilgamesh* trans. George, p. 86.

7 Ibid., p. 92.

8 Guillermo Algaze, 'Initial Social Complexity of Southwestern Asia: The Mesopotamian Advantage', *Current Anthropology* 42, 2 (April 2001), pp. 199–233.

9 Jean Bottéro, *Religion in Ancient Mesopotamia*, trans. Teresa Lavender Fagan (Chicago and London: Chicago University Press, 2001), p. 37.

10 Deuteronomy 11:6. [Translator's note: all biblical quotations use the New International Version.]

11 Deuteronomy 11:14.

12 Deuteronomy 11:17.

13 Deuteronomy 11:24.

14 Alain Testart, *La déesse et le grain: trois essais sur les religions néolithiques* (Paris: Errance, 2010), p. 18.

15 Ibid.

16 Ibid.

17 Daniel David Luckenbill, *Ancient Records of Assyria and Babylonia*, vol. II: *Historical Records of Assyria from Sargon to the End* (University of Chicago Press, 1927), p. 304.

18 Ibid.

19 Susan Wise Bauer, *The History of the Ancient World: From the Earliest Accounts to the Fall of Rome* (New York: W. W. Norton & Company, 2007), p. 410.

20 Nathan Haskell Dole (ed.), *The Works of J. W. von Goethe*, vol. IX, trans. Sir Walter Scott, Sir Theodore Martin, John Oxenford, Thomas Carlyle et al. (London and Boston: Francis A. Niccolls & Co., 1839), pp. 210–12.

21 Genesis 1:28.

22 *Gesenius's Hebrew and Chaldee Lexicon to the Old Testament Scriptures*, trans. Samuel Prideaux Tregelles (New York: John Wiley and Sons, 1890), p. 383.

23 Psalm 8:3–6.

24 Rashi on Genesis 1:28, at www.sefaria.org/Rashi_on_Genesis.1.1?lang=bi.

25 Percy Bysshe Shelley, *The Major Works*, ed. Zachary Leader and Michael O'Neill (Oxford and New York: Oxford University Press, 2003), p. 198.

26 Quoted in Tom Holland, *Dominion: The Making of the Western Mind* (London: Little, Brown, 2019), p. 155.

27 Saint Augustine, *On the Catechising of the Uninstructed*, trans. Revd S. D. F. Salmond, chs. 18 and 29, at www.ccel.org/ccel/schaff/npnf103.iv.iii.xix.html.

28 Saint Augustine, *On Genesis: Two Books on Genesis against the Manichees, and On the Literal Interpretation of Genesis: An Unfinished Book*, trans. Roland J. Teske (Washington: The Catholic University of America Press, 1991), p. 76.

29 Saint Augustine, *Against Faustus*, quoted in John Helgeland, Robert J. Daly and J. Patout Burns (eds.), *Christians and the Military: The Early Experience* (Philadelphia: Fortress Press, 1985), pp. 81f.

30 Saint Augustine, *The City of God*, trans. Marcus Dods (New York: Random House, 2010), p. 465.

31 Virgil, *The Aeneid*, trans. Cecil Day Lewis (Oxford and New York: Oxford University Press, 1998), book VI, p. 181.

32 Ibid.

33 Augustine, *City of God*, p. 444.

34 Ibid., p. 466.

35 Ibid., p. 441.

36 Ibid., p. 4.

37 Ibid., p. 441.

38 Holland, *Dominion*, p. xxii.

39 Homer, *The Odyssey*, trans. Walter Shewring (Oxford and New York: Oxford University Press, 1998), p. 100.

40 Quoted in Charles Freeman, *The Closing of the Western Mind* (London: Penguin, 2002), p. 299.

41 Ibid., p. 322.

42 Ibid., p. 242.

43 Ibid., p. 240.

44 Quoted in Joseph Carroll, *The Cultural Theory of Matthew Arnold* (Berkeley: University of California Press, 1982), p. 242 [translation modified].

45 Quoted in Hugo Bieber (ed.), *Heinrich Heine: A Biographical Anthology* (Philadelphia: The Jewish Publication Society of America, 1956), p. 263.

II Logos

1 W. H. Auden, 'Musée des Beaux Arts', in *Poems*, ed. Edward Mendelsohn, vol. I: *1927–1939* (Princeton University Press, 2022), pp. 338f.

2 Ovid, *Metamorphoses*, trans. Mary M. Innes (London: Penguin, 1955), p. 184.

3 Ibid., p. 185.

4 Girolamo Sernigi (1499), in Alvaro Velho, *A Journal of the First Voyage of Vasco da Gama, 1497–1499*, ed. E. G. Ravenstein (London: Hakluyt Society, 1898), p. 131.

5 Zhuangzi, *Basic Writings*, trans. Burton Watson (New York: Columbia University Press, 2003), p. 44.

6 Ibid., p. 32.

7 Zhuangzi, *The Complete Writings*, trans. Brook Ziporyn (Indianapolis: Hackett, 2020), p. 95.

8 Daniel R. Headrick, *Humans versus Nature: A Global Environmental History* (Oxford and New York: Oxford University Press, 2020), p. 138.

9 The dating is still a matter of some debate, and many researchers estimate a substantially longer period. I wrote at length about the Little Ice Age and its cultural effects in *Die Welt aus den Angeln* (Munich: Hanser, 2017).

10 René Descartes, *Discourse on the Method and the Meditations*, trans. John Veitch (New York: Cosimo Classics, 2008), pp. 45f.

11 Reproduced in Leonora Cohen Rosenfield, 'Descartes and Henry More on the Beast-Machine: A Translation of Their Correspondence Pertaining to Animal Automatism', *Annals of Science* 1 (1936), p. 50.

12 Ibid., p. 53.

13 Ibid., p. 52.

14 Michel de Montaigne, *The Complete Essays of Montaigne*, trans. Donald Murdoch Frame (Stanford University Press, 1965), pp. 330f.

15 Ibid., p. 331.

16 Ibid., p. 332.

17 Ibid., p. 333.

18 Ibid., p. 336.

19 Francis Bacon, *The New Organon*, ed. Lisa Jardine, Michael Silverthorne, trans. Michael Silverthorne (Cambridge University Press, 2000), p. xiv.

20 Bernardino Telesio, *De rerum natura iuxta propria principia*, book V, ch. III, vol. II, p. 216.

21 Bacon, *The New Organon*, p. 34.

22 Ibid., p. 70.

23 Ibid., p. 35.

24 Ibid., p. 40.

25 Ibid., p. 41.

26 Ibid., p. 43.

27 Ibid., p. 85.

28 Ibid., p. 53.

29 Ibid., p. 102.

30 Spinoza, *Ethics: Demonstrated in Geometric Order*, ed. Matthew J. Kisner, trans. Matthew J. Kisner and Michael Silverthorne (Cambridge University Press, 2018), p. 35.

31 Ibid., pp. 35f.

32 Ibid., p. 36.

33 Ibid., p. 37.

34 It was obvious that Spinoza would end up on the index, and he was in good company there, beside thinkers including Descartes, Montaigne, Bacon and Hobbes. The censors showed a keen instinct for historical significance.

35 Descartes, *Discourse on the Method and the Meditations*, p. 51.

36 For an exemplary and extensive discussion of this, see Dorothea Weltecke, *'Der Narr spricht: Es ist kein Gott': Atheismus, Unglauben und Glaubenszweifel vom 12. Jahrhundert bis zur Neuzeit* (Frankfurt: Campus, 2010).

37 Anthony Ashley Cooper, Earl of Shaftesbury, *An Inquiry Concerning Virtue, Or Merit* (1699), ed. David Walford (Manchester University Press, 1977), p. 11.

38 Soame Jenyns, *A Free Inquiry Into the Nature and Origin of Evil* (London: R. and J. Dodsley, 1758), p. 65.

39 Noël-Antoine Pluche, *Le Spectacle de la Nature, ou Entretiens sur ses Particularités de l'Histoire Naturelle*, vol. I (Paris: Veuve Estienne & Fils, 1752), p. iv.

40 Ibid., p. vii.

41 Ritchie Robertson, *The Enlightenment: The Pursuit of Happiness 1680–1790* (London: Penguin, 2020), p. 148.

42 Ibid., p. 188.

43 Immanuel Kant, *Universal Natural History and Theory of the Heavens*, in *Natural Science*, ed. Eric Watkins (Cambridge University Press, 2012), p. 270.

44 Ibid.

45 Voltaire, 'Poem on the Lisbon Disaster', in *Toleration and Other Essays*, ed. and trans. John McCabe (New York and London: G. P. Putnam's Sons, 1912), p. 255.

46 Johann Gottfried Herder, *Ideas for the Philosophy of the History of Mankind*, trans. Gregory Martin Moore (Princeton University Press, 2024), pp. 15f.

47 Kant, *Universal Natural History*, pp. 194f.

48 Immanuel Kant, 'What Is Enlightenment', in *Political Writings*, ed. H. S. Reiss, trans. H. B. Nisbet (Cambridge University Press, 1991), p. 55.

49 Peter Gay, *The Enlightenment: An Interpretation*, vol. II: *The Science of Freedom* (New York: Knopf, 1969), p. 368.

50 Quoted in David Day, *Conquest: How Societies Overwhelm Others* (Oxford and New York: Oxford University Press, 2013), p. 159.

51 Thomas Jefferson, *Notes on the State of Virginia* (1802), ed. William Harwood Peden (New York: W. W. Norton, 1982), p. 130.

52 Ibid.

53 Jonathan Swift, *Gulliver's Travels* (Oxford and New York: Oxford University Press, 2008), p. xxiii.

54 Paul Henry Thiry d'Holbach, *The System of Nature: Or, the Laws of the Physical and Moral World* (London: Davison, 1820), p. v.

55 Ibid., pp. 1f.

56 Ibid., pp. 149f.

57 Ibid., pp. 170f.

58 Denis Diderot, *D'Alembert's Dream*, in *Rameau's Nephew and Other Works*, trans. Jacques Barzun and Ralph Henry Bowen (Indianapolis: Hackett, 2001), pp. 133f.

59 Letter of 15 October 1759, in Denis Diderot, *Diderot's Letters to Sophie Volland: A Selection*, trans. Peter France (Oxford and New York: Oxford University Press, 1972), p. 38.

60 Jean-Jacques Rousseau, *Discourse on the Origin of Inequality*, trans. Donald A. Cress (Indianapolis: Hackett, 1992), p. 19.

61 Ibid., p. 25.

62 Ibid., p. 44.

63 Ibid.

64 Jeremy Bentham, *A Comment on the Commentaries and a Fragment on Government*, ed. J. H. Burns and H. L. A. Hart (Oxford: Clarendon Press, 2008), p. 393.

65 Jeremy Bentham, *An Introduction to the Principles of Morals and Legislation*, ed. J. H. Burns and H. L. A. Hart (Oxford: Clarendon Press, 1996), p. 11.

66 Ibid., p. 311.

67 [Translator's note: 'Even if it is not true, it is a good invention', a saying attributed to Giordano Bruno.]

68 Jeremy Bentham, *Panopticon: or, The Inspection-House* (Dublin: Thomas Byrne, 1791), p. iii.

69 Ibid., p. 2.

70 A different version of this section on Bentham was published in 2017 in my essay *Gefangen im Panoptikum* (Vienna: Residenz Verlag) [Imprisoned in the Panopticon].

71 Carl Schmitt, *The Nomos of the Earth*, trans. G. Ulmen (New York: Telos, 2003), p. 47.

72 Ernest Renan, 'The Intellectual and Moral Reform of France' (1871), in *What Is a Nation? And Other Political Writings*, ed. and trans. M. F. N. Giglioli (New York: Columbia University Press, 2018), p. 232.

73 Carl Hagenbeck, *Von Tieren und Menschen* (Leipzig: Paul List, 1967), p. 4. [Translator's note: this passage is not contained in the English edition.]

74 Carl Hagenbeck, *Beasts and Men*, trans. (abridged) Hugh S. R. Eliot and A. G. Thacker (London: Longmans, Green and Co., 1912), p. 155.

75 Hagenbeck, *Von Tieren und Menschen*, p. 43.

76 Ibid., p. 57.

77 Ibid., p. 61.

78 Thomas Phillips, *A Journal of a Voyage Made in the Hannibal of London, Ann. 1693, 1694, From England, to Cape Monseradoe, in Africa, And thence along the Coast of Guiney to Whidaw, the Island of St. Thomas, And so forward to Barbadoes* (London: Walthoe, 1732), p. 196.

79 Ibid., p. 219.

80 Ibid.

81 Ibid., p. 236.

82 Quoted in George Francis Dow, *Slave Ships and Slaving* (Mineola: Dover, 2002), p. 89.

83 Ibid., p. 106.

84 Ibid., p. xxi.

85 Ibid.

86 Ibid., p. xxiii.

87 Ibid., p. xxiv.

88 Jefferson, *Notes on the State of Virginia*, p. 264.

89 Ibid.

90 G. W. F. Hegel, *The Philosophy of History*, trans. J. Sibree (Mineola: Dover, 2002), p. 98.

91 Ibid., p. 96.

92 Ibid., p. 95.

93 Ibid., p. 86.

94 Ibid., p. 81.

95 Ibid., p. 84.

96 Ibid., p. 82.

97 Ibid.

98 Ibid., p. 95.

99 Schmitt, *The Nomos of the Earth*, pp. 198f.

100 Immanuel Kant, 'Observations on the Feeling of the Beautiful and Sublime', in *Observations on the Feeling of the Beautiful and Sublime and Other Writings*, ed. Patrick Frierson and Paul Guyer, trans. Patrick Frierson (Cambridge University Press, 2011), p. 61.

101 Immanuel Kant, 'On the Use of Teleological Principles in Philosophy', quoted in Robert Bernasconi, 'Kant as an Unfamiliar Source of Racism', in *Philosophers on Race: Critical Essays*, ed. Julie K. Ward and Tommy L. Ott (Oxford: Blackwell, 2002), p. 158.

102 Ibid., p. 148.

103 Voltaire, *Traité de metaphysique*, ed. H. T. Patterson (Manchester University Press, 1937), p. 33.

104 Georges Baron Cuvier, *The Animal Kingdom Arranged in Conformity with Its Organization*, ed. and trans. Henry M'Murtrie (New York: G. & C. & H. Carvill, 1832), p. 50.

105 Charles Darwin, *The Descent of Man, and Selection in Relation to Sex* (Princeton University Press, 1981), p. 405.

106 Francis Galton, *The Narrative of an Explorer in Tropical South Africa* (London: John Murray, 1853), pp. 123f.

107 Francis Galton, 'Photographic Composites', *The Photographic News* (1885), p. 243.

108 Ibid.

109 Revd Joseph S. Exell and Henry Donald Maurice Spence-Jones, *The Pulpit Commentary*, 48 vols. (London: Funk & Wagnalls Company, 1880–97), vol. I.

110 Pankaj Mishra, *Age of Anger: A History of the Present* (London: Penguin, 2017), p. 48.

111 Hervé Faye, *Sur l'origine du monde* (Paris: Gauthier-Villars, 1884), p. 110.

112 Friedrich Nietzsche, *The Gay Science*, trans. Josefine Nauckhoff (Cambridge University Press, 2001), p. 120.

113 Sigmund Freud, *Civilization and Its Discontents*, trans. James Strachey (London and New York: Norton, 1961), p. 84.

114 Ibid., p. 95.

115 Nietzsche, *The Gay Science*, p. 201.

116 Ibid., p. 210.

117 Ibid., p. 221.

118 Ibid., p. 241.

119 Samuel Taylor Coleridge, 'The Rime of the Ancient Mariner', in *Coleridge: Poems* (New York: Knopf Doubleday, 2014), p. 50.

120 William Blake, 'Auguries of Innocence', in *Collected Poems*, ed. W. B. Yeats (London and New York: Routledge, 2002), p. 88.

III Cosmos

1 Quoted in Headrick, *Humans versus Nature*, p. 1.

2 Leon Trotsky, *Literature and Revolution*, trans. Rose Strunsky (Chicago: Haymarket, 2005), pp. 204f.

3 Headrick, *Humans versus Nature*, p. 315.

4 Elizabeth Economy, *The River Runs Black: The Environmental Challenge to China's Future* (Ithaca: Cornell University Press, 2004), p. 49.

5 Nietzsche, *The Gay Science*, p. 213.

6 Ibid., pp. 247f.

7 Montaigne, 'An Apology for Raymond Sebond', in *The Complete Essays of Montaigne*, pp. 328f.

8 Ibid., p. 329.

9 Ibid., p. 330.

10 Ibid., p. 337.

11 Alexander von Humboldt, *Views of the Cordilleras and Monuments of the Indigenous Peoples of the Americas*, ed. Vera M. Kutzinski and Ottmar Ette, trans. J. Ryan Poynter (Chicago and London: Chicago University Press, 2013), p. 6.

12 Ibid., pp. 5f.

13 Ibid., p. 11.

14 Pliny the Elder, *Natural History*, trans. John F. Healy (London: Penguin, 1991), p. 75.

15 Alexander von Humboldt, *Cosmos: A Sketch of a Physical Description of the Universe*, vol. I, trans. Elise C. Otté (London: Henry G. Bohn, 1848), pp. 2f.

16 Friedrich Nietzsche, *Thus Spoke Zarathustra*, trans. Graham Parkes (Oxford and New York: Oxford University Press, 2008), p. 36.

17 Bruno Latour and Peter Weibel, *Critical Zones: The Science and Politics of Landing on Earth* (Cambridge, MA: MIT Press, 2020), p. 13.

18 James E. Lovelock, *Reviews of Geophysics* 17 (11 May 1989), pp. 215–22.

19 Latour and Weibel, *Critical Zones*, p. 5.

20 Ibid.

21 Ibid.

22 Latour, *Facing Gaia*, p. 36.

23 Hartmut Rosa, *Resonance: A Sociology of Our Relationship to the World*, trans. James C. Wagner (Cambridge: Polity, 2019).

24 In *Neue Zürcher Zeitung*, 18 December 2021: www.nzz.ch/feuilleton/alexander -kluge-zeichnet-eine-grosse-beobachtungsgabe-aus-ld.1660027.

Index

subjugation of nature (*cont.*)
  victims and consequences of 203–4
  as a voluntary act 218–19
  Western tradition xviii–xix
  white male 203
Suleiman the Magnificent 100–1
Sumerians 12, 13, 14, 18–19
  clay tablets 19
  mythology 48
superego 207–8
Swift, Jonathan 158
The System of Nature (d'Holbach) 162

Talos 82
Tao/Taoism xix, 95
Taqi ad-Din Muhammad ibn Ma'ruf
  102–3
taxation 13, 17, 29
Telesio, Bernardino 123–5, 127, 129, 248
Tenochtitlan 14, 15
testaceotheology (snails) 147
Testament of Jean Meslier (Meslier) 161
Testart, Alain 25
theology 66, 106, 115, 120, 130, 155, 229
  during the Enlightenment 152–3,
    155–6, 159–60, 176
  freedom of will 156
  Kant 154–5
  legacy of 233–4
  theological thinking 276
  theological tradition 276
  see also physico-theology
Theresia, Empress Maria 244
Theseus 31
Thirty Years War 102, 109
Three Gorges Dam 224
Tiananmen Square xiii
totalitarianism 237
Tower of Babel 47
trade 25, 101, 183
trade routes 19
Traité de mécanique céleste (Treatise on
  Celestial Mechanics, Laplace) 204

trees 273
triangular trade 183, 190–1, 210
tribal societies 167
Trinity Test (1945) 232
Trotsky, Leon 224
tsunami 149–50
Turner, Joseph Mallard William 212–13,
  214–15, 219
two-substance doctrine 117, 118–20, 124

United States of America (USA) 157
*Universal Natural History and Theory of
  the Heavens* (Kant) 150, 153, 153–4
universe 150, 153
  as clockwork 143, 156
  interconnected systems 248
  mechanical 242–3
  system of subordination 146
  see also world
University College in London 174–5
urban bourgeoisie 175
urban cultures 13, 14, 232
Uruk 6–8, 9, 11, 12, 14, 16, 17
Uruk vase 3–5, 18, 30
Uta-napishti 8–9, 10, 47
utilitarianism 170, 171

va-kibsu-ha ('and you shall subdue it')
  49–50
vacuum pumps 137, 138
Venus figurines 25, 33
Venus of Willendorf 23–6, 35
Verdi, Giuseppe 208
Versailles 132–4
Victorian bourgeoisie 205–6
violence 41, 60, 69
  in ancient Greece 70
  and Christianity 69, 70, 71, 72
  justification of 69, 70, 71–2, 106, 107
Vira Alakesvara, King 91–2
Virgil 62
virtue 164
Volga 224